FRONTIERS OF MODEL PREDICTIVE CONTROL

Edited by **Tao Zheng**

Frontiers of Model Predictive Control
Edited by Tao Zheng

Published by InTech
Janeza Trdine 9, 51000 Rijeka, Croatia

Publishing Process Manager Oliver Kurelic

Technical Editor Teodora Smiljanic

Cover Designer InTech Design Team

Additional hard copies can be obtained from orders@intechopen.com

Frontiers of Model Predictive Control, Edited by Tao Zheng
p. cm.
ISBN 978-953-51-0119-2

For EU product safety concerns:
IN TECH d.o.o., Prolaz Marije Krucifikse Kozulić 3, 51000 Rijeka, Croatia, info@intechopen.com or visit our website at intechopen.com.

Meet the editor

Dr Tao Zheng was born in Chongqing, China, in 1981. In 2003 and 2008, he received his BE degree and PhD degree, respectively, from University of Science and Technology of China, both in Automatic Control. He was also awarded the President's Excellence Prize by the Chinese Academy of Sciences in 2008. He currently holds the Associate Researcher (equivalent to Associate Professor) position with the Department of Automation, Hefei University of Technology, China. Dr Zheng's recent interests are in Advanced Model Predictive Control (including Nonlinear Model Predictive Control and Multi-Objective Model Predictive Control), Multi-Objective Optimization and Multi-objective Genetic Algorithm. He is also interested in the application of optimization theory to economy, for example the optimization problem in currency-circulation system.

Contents

Preface

Model Predictive Control (MPC) is not only the name for a special kind of control algorithms or controllers based on these algorithms, but also the name for a powerful thought in control theory. MPC has a very special originating process; because its industrial application appeared much earlier than its theoretical description, it could solve process control problems without precise theory. However, from the perspective of its essence, this special originating process indicates high coherence between MPC and natural thinking manner of humans, bringing out control problems to be solved. To first predict using a model, and then provide a control law, considering the predicted result, seems to be the most comprehensible, convenient and understandable way in controlling.

However, limited by modeling and optimization method during the early days, MPC could only be used in process industry, using a local linear model and a large sample period. In fact, this poor situation lasted for decades following MPC's birth. In some researchers' minds, MPC still remains a different name for advanced process control. The advantages of MPC were underestimated seriously during those years

While we are aware that the real world is much more complex than the linear, time-invariant, lumped-parameter and deterministic model in classical MPC, if we want to have better control performance, these factors must be taken into account. The advantages of MPC can only be demonstrated entirely and clearly if MPC can handle much more kinds of system models.

Fortunately, during the recent years, the rapid development of computational science and technology led to the second "boom" of MPC. Nonlinearity, stochastic character, robustness and many other factors started to be considered. Efficient applicable MPC algorithms have been established using modern computational techniques, such as the genetic algorithm. Theoretical achievements have also been obtained with the efforts of many control scientists. Applications of MPC can now be found in almost all engineering domains.

To start with, this book will introduce the basic structure and the historical development of MPC, for readers who are not so familiar with the topic. Some distinctive examples of recent MPC use will then be presented, both in the theoretical and the engineering domain, to illustrate the frontiers of this field. This special structure can help the readers who want to acquaint themselves with MPC in general, while readers who want to study MPC in one particular direction can also get helpful guidance in this book.

The book's authors from around the world appreciate the contributions made by researchers before them, and bear in mind the quote 'we stand on the shoulders of giants'. We would also like to thank all the people who helped us greatly in the writing process, including our colleagues and friends, and especially the zealous managers and editors at InTech. Finally, we thank all of our family members, you are always our ultimate love and help.

Tao Zheng
Hefei University of Technology,
China

Introductory Chapter

Model Predictive Control: Basic Characters

Tao Zheng
Hefei University of Technology, China

1. Introduction

The name 'Model predictive control' exactly indicates the three most essential characters of this kind of controllers, a model can be used to predict the future behaviour of the system, the prediction based on above model and historical data of the system and online optimal control based on above prediction and certain control criterion.

2. The predictive model

Any model that could be used to predict the future behaviour can be the system model in MPC, and it is usually called predictive model.

MPC itself has no special request on the choice of model, the only need is that the model could predict the future behaviour of the system, no matter how we get the system model and how we obtain the future output by the model. But many researchers still classify MPC into different types by their models, since different model usually lead to quite different optimization method in solution of control law. Because all MPC have the same basic structure, the optimization method may be the most important part of a novel MPC algorithm indeed, and it also can determine the algorithm's practical applicability in industry. In Certain Meaning, the develop history of MPC is mainly the develop history of the predictive model of MPC.

When MPC was invented in 1970s, limited by the modelling and computational method, the scientist and engineers often use simple models, such as discrete time linear model (Richalet *et al.*, 1978, Culter *et al.*, 1980, Rouhani *et al.*, 1982 and Clarke *et al.*, 1987), to build MPC, while using this kind of models could already satisfy the requirement on control performance in process industry of that days. Later, based on modern control theory, a lot of MPC based on linear state-space system model is proposed (Ordys *et al.*, 1993, Lee *et al.*, 1994). These mentioned references can also help the readers of this book to understand the basic characters thoroughly if they still have problems after reading this short guidance, because these references were work of the precursors, who paid special attention to explain what MPC's essential properties are.

But, nonlinearity, constraints, stochastic characters and other complex factors exist naturally in the physical world, especially in control engineering.

For highly nonlinear processes, and for some moderately nonlinear processes, which have large operating regions, MPC based on local linear model is often inefficient. Since the nonlinearity is the most important essential nature, and the increasing demand on the control performances, controller designers and operators have to face it directly. In 1990s, nonlinear model predictive control (NMPC) became one of the focuses of MPC research and it is still difficult to handle today as Prof. Qin mentioned in his survey (Qin *et al.*, 2003). The direct incorporation of a nonlinear process into the MPC formulation will result in a non-convex nonlinear programming problem, which needs to be solved under strict sampling time constraints. In general, there is still no analytical solution to this kind of nonlinear programming problem. To solve this difficulty, many kinds of simplified model is chosen to present nonlinear systems, such as nonlinear affine model (Cannon, 2004), bilinear model (Yang *et al.*, 2007), block-oriented model (including Hammerstein model, Wiener model, *etc.*)(Harnischmacher *et al.*, 2007, Arefi *et al.*, 2008).

Stochastic characters and other complex factors also special expression models, such as Markov chain description and other method. Limited by the length, we won't introduce them in detail here, readers who are interested in these models can read more surveys on MPC and then find clue to research on them.

3. The prediction

In Fig. 1., the basic principle of MPC is illustrated. It is also very convenient to explain the term 'Prediction' in MPC.

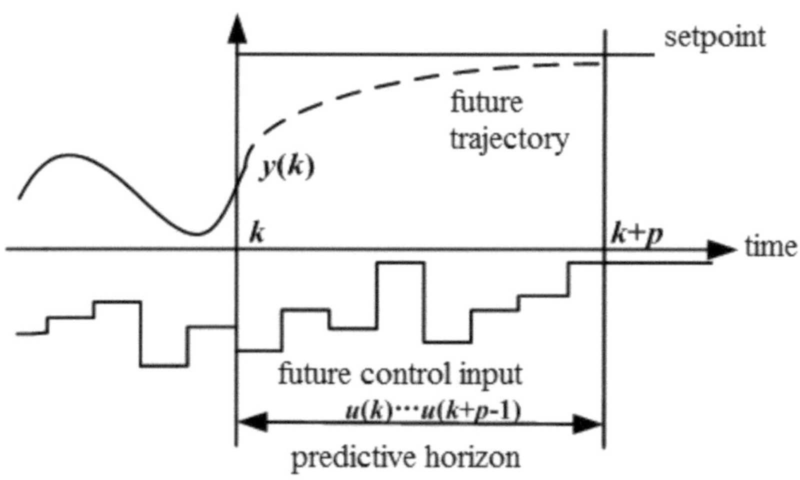

Fig. 1. Basic principle of Model Predictive Control

Consider a SISO discrete system for example, with integer k representing the current discrete time, $y(k)$ representing output and $u(k)$ representing control input. At time k, the historic output $y(k\text{-}1)$, $y(k\text{-}2)$, $y(k\text{-}3)$ …, historic control input $u(k\text{-}1)$, $u(k\text{-}2)$, $u(k\text{-}3)$…and the instant output $y(k)$ are known, if we also know the value of instant control input $u(k)$, the next future output $y(k+1|k)$ can be predicted. This operation is usually called as one-step prediction.

With similar process, if we know the sequence of future control input $u(k)$, $u(k+1)$, $u(k+2)$, $u(k+3)$..., we can predict the sequence of future output $y(k+1|k)$, $y(k+2|k)$, $y(k+3|k)$ ..., here, the length of prediction or the number of predictive steps is called predictive horizon in MPC.

In MPC, though we cannot know he sequence of future control input $u(k)$, $u(k+1)$, $u(k+2)$, $u(k+3)$..., we can still predict $y(k+1|k)$, $y(k+2|k)$, $y(k+3|k)$, with he sequence of future control input $u(k)$, $u(k+1)$, $u(k+2)$, $u(k+3)$... remaining in these predictive values as unknown variables that need to be solved.

If certain expectation future output is given, such as the future trajectory shown in Fig. 1. (the expect way of output how it reaches the setpoint in certain time), to the contrary of prediction mentioned in the second and the third paragraph of this section, the sequence of future control input $u(k)$, $u(k+1)$, $u(k+2)$, $u(k+3)$... can be solved by the given $y(k+1|k)$, $y(k+2|k)$, $y(k+3|k)$ ..., and this is exactly the way how MPC can get a optimal control law from model prediction.

4. The online optimal control law

If a future output trajectory or an objective faction (usually a quadratic function of input and output) is given, in MPC, as mentioned above, the optimal control law can be solved.

At time k, a sequence of future input will be solved $u(k)$, $u(k+1)$, $u(k+2)$, $u(k+3)$, but only instant input $u(k)$ will be carry out actually by the system. At the next sample time, time $k+1$, the whole process of prediction and optimization will be repeated and a new future input sequence is obtained. This is the essence of online optimization.

This operation can introduce information into the controller, such as the error between predictive output and real output, so model mismatch and other disturbances can be eliminated gradually. In some extent, the online optimization can be recognized as a kind of feedback control.

For linear systems, control law of MPC can often be obtained analytically, but for most nonlinear systems, we have to use numerical optimization algorithms to get the control solution. Nowadays modern numerical optimization methods, such as Genetic Algorithm (GA) (Yuzgec *et al.*, 2006), ant colony optimization (ACO), Particle Swarm Optimization (PSO) *etc.* are the common solution tool for NMPC.

Compared to MPC for SISO system, for MIMO system or multi-objective problem, there is no special difference in optimization methods. While, constraints (on input, on output or on both of them) may cause big trouble in online optimization, for linear system, there is some method that can deal with simple constraints, but for complex constraints or for nonlinear systems, numerical methods are still the only usable means.

5. Application of MPC

When MPC is invented, limited by modeling and optimization method and tools, it could only be used in process industry, with local linear model and large sample period. And the position of MPC in a whole process control project is shown in Fig. 2. We can see that, MPC is in a 'middle' level.

Now, the rapid development in computational science and technology leads to the second boom of MPC, especially on the applicative research of it. MPC's application can be found almost in every engineering field rather than process industry, such as MPC in motion control (Richalet, 1993), modern agriculture (Coelho *et al.*, 2005), communication (Chisci *et al.*, 2006) and even in decision making science (Kouvaritakis *et al.*, 2006). In this book, there are also several recent successful applicative example of MPC for interesting plant for you.

It can be believed with much confidence, in the future, the great benefit of MPC could be shared by more and more practical domain for more and more people in the world.

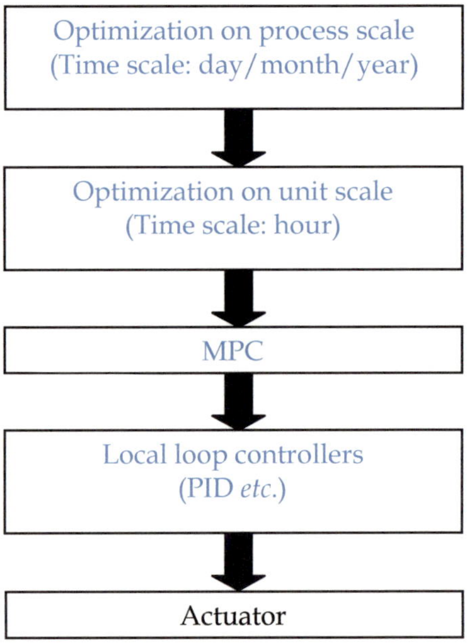

Fig. 2. Position of MPC in a typical process control project

6. Acknowledgement

The author thanks the help from teachers, colleagues and friends, especially, Professor Gang WU from University of Science and technology of China, Associate Professor Wei CHEN from Hefei University of Technology and Associate Professor De-Feng HE from Zhejiang University of Technology.

This work is supported Special Foundation for Ph. D. of Hefei University of Technology (No. 2010HGBZ0616) and Inventive Project for Young Teachers of Hefei University of Technology (2011HGQC0994), both comes from the Fundamental Research Funds for the Central Universities, China.

7. References

Arefi M. M.; Montazeri A.; Poshtan J.; Jahed-Motlagh M. R. (2008). Wiener-neural identification and predictive control of a more realistic plug-flow tubular reactor. *Chemical Engineering Journal*, Vol.138, No.1-3, May, 2008, pp.274-282, ISSN 1385-8947.

Cannon M. (2004). Efficient nonlinear model predictive control algorithms. *Annual Reviews in Control*, Vol.28, No.2, 2004, pp.229-237, ISSN 1367-5788.

Chisci L.; Pecorella T.; Fantaccim R. (2006). Dynamic bandwidth allocation in GEO satellite networks: a predictive control approach. *Control Engineering Practice*, September, 2006, VOl.14, No.9, pp.1057-1067, ISSN 0967-0661.

Clarke D. W.; Montadi C.; Tuffs P. S. (1987). Generalized predictive control. *Automatica*, Vol.23, No.2, March, 1987, pp.137-162, ISSN 0005-1098.

Coelho J. P; Moura Oliveira P. B. de; Cunha J. B. (2005). Greenhouse air temperature predictive control using the particle swarm optimization algorithm. *Computers and Electronics in Agriculture*, December, 2005, Vol.49, No.3, pp.330-344, ISSN 0168-1699.

Cutler C. R.; Ramaker B. L. (1980). Dynamic matrix control: a computer control algorithm. Proceedings of the Joint Automatic Control Conference. San Francisco, 1980, Vol. 1, WP5-B.

Harnischmacher G.; Marquardt W. (2007). Nonlinear model predictive control of multivariable processes using block-structured models. *Control Engineering Practice*, Vol.15, No.10, October, 2007, pp.1238-1256, ISSN 0967-0661.

Kouvaritakis B.; Cannon M.; Couchman P. (2006). MPC as a tool for sustainable development integrated policy assessment. IEEE Transactions on Automatic Control, January, 2006, Vol.51, No.1, pp.145-149, ISSN 0018-9286.

Lee J. H.; Morari M.; García C. E. (1994). State space interpretation of model predictive control. *Automatica*, Vol.30, No.4, April, 1994, pp.707-717, ISSN 0005-1098.

Ordys A. W.; Clarke D. W. (1993). A state-space description for GPC controllers. *International Journal of System Science*, Vol.24, No.9, September, 1993, pp.1727-1744, ISSN 0020-7721.

Qin S. J; Badgwell T. A. (2003). A survey of industrial predictive control technology. *Control Engineering Practice*, Vol.11, No.7, July, 2003, pp.733-764, ISSN 0967-0661.

Richalet J. (1993). Industrial applications of model based predictive control. *Automatica*, Vol.29, No.5, July, 1993, pp.1251-1274, ISSN 0005-1098.

Richalet J.; Rault A.; Testud J. L.; Papon J. (1978). Model predictive heuristic control: applications to industrial processes. *Automatica*, Vol.14, No.5, September, 1978, pp.413-428, ISSN 0005-1098.

Rouhani.; R, Mehra R. K. (1982). Model algorithmic control: basic theoretical properties. *Automatica*, Vol.18, No.4, July, 1982. pp.401-414, ISSN 0005-1098.

Yang H.; Li S Y. (2007). A date-driven bilinear predictive controller design based on subspace method. Proceedings of IEEE International Conference on Control Applications, 2007, Singapore, pp.176-181.

Yuzgec U.; Becerikli Y.; Turker M. (2006). Nonlinear predictive control of a drying process using genetic algorithms. *ISA Transactions*, Vol.45, No.4, October, 2006, pp.589-602, ISSN 0019-0578.

Part 1

New Theoretical Frontier

1

A Real-Time Gradient Method for Nonlinear Model Predictive Control

Knut Graichen and Bartosz Käpernick
Institute of Measurement, Control and Microtechnology
University of Ulm
Germany

1. Introduction

Model predictive control (MPC) is a modern control scheme that relies on the solution of an optimal control problem (OCP) on a receding horizon. MPC schemes have been developed in various formulations (regarding continuous/discrete-time systems, finite/infinite horizon length, terminal set/equality constraints, etc.). Comprehensive overviews and references on MPC can, for instance, be found in Diehl et al. (2009); Grüne & Pannek (2011); Kothare & Morari (2000); Mayne et al. (2000); Rawlings & Mayne (2009).

Although the methodology of MPC is naturally suited to handle constraints and multiple-input systems, the iterative solution of the underlying OCP is in general computationally expensive. An intuitive approach to reducing the computational load is to solve the OCP approximately, for instance, by using a fixed number of iterations in each sampling step. In the next MPC step, the previous solution can be used for a warm-start of the optimization algorithm in order to successively reduce the suboptimality of the predicted trajectories. This incremental strategy differs from the "optimal" MPC case where the (numerically exact) OCP solution is assumed to be known.

There exist various suboptimal and real-time approaches in the literature with different kinds of terminal constraints and demands on the optimization algorithm (Cannon & Kouvaritakis, 2002; DeHaan & Guay, 2007; Diehl et al., 2005; Graichen & Kugi, 2010; Lee et al., 2002; Michalska & Mayne, 1993; Ohtsuka, 2004; Scokaert et al., 1999). In particular, the approaches of Ohtsuka (2004) and Diehl et al. (2005) are related to the MPC scheme presented in this chapter. In Ohtsuka (2004), an algorithm is developed that traces the solution of the discretized optimality conditions over the single sampling steps. The real-time iteration scheme presented by Diehl et al. (2005) uses a Newton scheme together with terminal constraints in order to compute an approximate solution that is refined in each sampling step.

Suboptimal MPC schemes require special attention regarding their convergence and stability properties. This is particularly important if an MPC formulation without terminal constraints is used in order to minimize the computational complexity and to allow for a real-time implementation for very fast dynamical systems. In this context, a suboptimal MPC approach without terminal constraints was investigated in Graichen & Kugi (2010). Starting from the assumption that an optimization algorithm with a linear rate of convergence exists, it is shown that exponential stability of the closed-loop system as well as exponential decay of the suboptimality can be guaranteed if the number of iterations per sampling step satisfies a lower bound (Graichen & Kugi, 2010). The decay of the suboptimality also illustrates the incremental improvement of the MPC scheme.

Based on these theoretical considerations (Graichen & Kugi, 2010), this chapter presents a real-time MPC scheme that relies on the gradient method in optimal control (Dunn, 1996; Graichen et al., 2010; Nikol'skii, 2007). This algorithm is particularly suited for a real-time implementation, as it takes full advantage of the MPC formulation without terminal constraints. In addition, the gradient method allows for a memory and time efficient computation of the single iterations, which is of importance in order to employ the MPC scheme for fast dynamical systems.

In this chapter, the gradient-based MPC algorithm is described for continuous-time nonlinear systems subject to control constraints. Starting from the general formulation of the MPC problem, the stability properties in the optimal MPC case are summarized before the suboptimal MPC strategy is discussed. As a starting point for the derivation of the gradient method, the necessary optimality conditions for the underlying OCP formulation without terminal constraints are derived from Pontryagin's Maximum Principle. Based on the optimality conditions, the gradient algorithm is described and its particular implementation within a real-time MPC scheme is detailed. The algorithm as well as its properties and incremental improvement in the MPC scheme are numerically investigated for the double pendulum on a cart, which is a benchmark in nonlinear control. The simulation results as well as the CPU time requirements reveal the efficiency of the gradient-based MPC scheme.

2. MPC formulation

We consider a nonlinear continuous-time system of the form

$$\dot{x}(t) = f(x(t), u(t)) \quad x(t_0) = x_0 \,, \quad t \geq t_0 \tag{1}$$

with the state $x \in \mathbb{R}^n$ and the control $u \in \mathbb{R}^m$ subject to the control constraints

$$u(t) \in [u^-, u^+] \,. \tag{2}$$

Without loss of generality, we assume that the origin is an equilibrium of the system (1) with $f(0,0) = 0$. Moreover, the system function f is supposed to be continuously differentiable in its arguments. This section summarizes the MPC formulation as well as basic assumptions and basic results for the stability of the MPC scheme in closed-loop.

2.1 Optimal control problem

For stabilizing the origin of the system (1), an MPC scheme based on the following optimal control problem (OCP) is used

$$\min_{\bar{u} \in \mathcal{U}_{[0,T]}} \quad J(x_k, \bar{u}) = V(\bar{x}(T)) + \int_0^T l(\bar{x}(\tau), \bar{u}(\tau)) \, \mathrm{d}\tau \tag{3a}$$

$$\text{s.t.} \quad \dot{\bar{x}}(\tau) = f(\bar{x}(\tau), \bar{u}(\tau)) \,, \quad \bar{x}(0) = x_k = x(t_k) \,, \tag{3b}$$

where $\mathcal{U}_{[0,T]}$ is the admissible input space

$$\mathcal{U}_{[0,T]} := \{u(\cdot) \in L_\infty^m[0,T] : u(t) \in [u^-, u^+],\, t \in [0,T]\}\,. \tag{4}$$

The initial condition $x(t_k) = x_k$ in (3b) denotes the measured (or observed) state of the system (1) at time $t_k = t_0 + k\Delta t$ with the sampling time Δt. The bared variables $\bar{x}(\tau)$, $\bar{u}(\tau)$ represent internal variables of the controller with the MPC prediction time coordinate $\tau \in [0,T]$ and the horizon length $T \geq \Delta t$.

The integral and the terminal cost functions in (3a) are assumed to be continuously differentiable and to satisfy the quadratic bounds

$$\begin{aligned} m_l(||x||^2 + ||u||^2) \leq & l(x,u) \leq M_l(||x||^2 + ||u||^2) \\ m_V||x||^2 \leq & \; V(x) \; \leq M_V||x||^2 \end{aligned} \tag{5}$$

for some constants $m_l, M_l > 0$ and $m_V, M_V > 0$. The optimal solution of OCP (3) is denoted by

$$\bar{u}_k^*(\tau) := \bar{u}^*(\tau; x_k), \quad \bar{x}_k^*(\tau) := \bar{x}^*(\tau; x_k, \bar{u}_k^*), \quad \tau \in [0,T], \quad J^*(x_k) := J(x_k, \bar{u}_k^*)\,. \tag{6}$$

To obtain a stabilizing MPC feedback law on the sampling interval $[t_k, t_{k+1})$, the first part of the optimal control $\bar{u}_k^*(\tau)$ is used as control input for the system (1)

$$u(t_k + \tau) = \bar{u}_k^*(\tau) =: \kappa(\bar{x}_k^*(\tau); x_k)\,, \quad \tau \in [0, \Delta t)\,, \tag{7}$$

which can be interpreted as a nonlinear "sampled" control law with $\kappa(0; x_k) = 0$. In the next MPC step at time t_{k+1}, OCP (3) is solved again with the new initial condition x_{k+1}. In the absence of model errors and disturbances, the next point x_{k+1} is given by $x_{k+1} = \bar{x}_k^*(\Delta t)$ and the closed-loop trajectories are

$$\begin{aligned} x(t) &= x(t_k + \tau) = \bar{x}^*(\tau; x_k)\,, \\ u(t) &= u(t_k + \tau) = \bar{u}^*(\tau; x_k)\,, \quad \tau \in [0, \Delta t)\,, \quad k \in \mathbb{N}_0^+\,. \end{aligned} \tag{8}$$

2.2 Domain of attraction and stability

The following lines summarize important results for the "optimal" MPC case without terminal constraints, i.e. when the optimal solution (6) of OCP (3) is assumed to be known in each sampling step. These results are the basis for the suboptimal MPC case treated in Section 3. Some basic assumptions are necessary to proceed:

Assumption 1. *For every* $x_0 \in \mathbb{R}^n$ *and* $u \in \mathcal{U}_{[0,T]}$, *the system* (1) *has a bounded solution over* $[0,T]$.

Assumption 2. *OCP* (3) *has an optimal solution* (6) *for all* $x_k \in \mathbb{R}^n$.

Since u is constrained, Assumption 1 is always satisfied for systems without finite escape time. Moreover, note that the existence of a solution of OCP (3) in Assumption 2 is not very restrictive as no terminal constraints are considered and all functions are assumed to be continuously differentiable.[1].

---

[1] Theorems on existence and uniqueness of solutions for certain classes of OCPs can, for instance, be found in Berkovitz (1974); Lee & Markus (1967).

An MPC formulation without terminal constraints has been subject of research by several authors, see for instance Graichen & Kugi (2010); Ito & Kunisch (2002); Jadbabaie et al. (2001); Limon et al. (2006); Parisini & Zoppoli (1995). Instead of imposing a terminal constraint, it is often assumed that the terminal cost V represents a (local) Control Lyapunov Function (CLF) on an invariant set S_β containing the origin.

Assumption 3. *There exists a compact non-empty set* $S_\beta = \{x \in \mathbb{R}^n : V(x) \leq \beta\}$ *and a (local) feedback law* $q(x) \in [u^-, u^+]$ *such that* $\forall x \in S_\beta$

$$\frac{\partial V}{\partial x} f(x, q(x)) + l(x, q(x)) \leq 0 \,. \tag{9}$$

There exist several approaches in the literature for constructing a CLF as terminal cost, for instance Chen & Allgöwer (1998); Primbs (1999). In particular, $V(x)$ can be designed as a quadratic function $V(x) = x^{\mathsf{T}} P x$ with the symmetric and positive definite matrix P following from a Lyapunov or Riccati equation provided that the linearization of the system (1) about the origin is stabilizable.

An important requirement for the stability of an MPC scheme without terminal constraints is to ensure that the endpoint of the optimal state trajectory $\bar{x}_k^*(T)$ reaches the CLF region S_β. The following theorem states this property more clearly and relates it to the overall stability of the (optimal) MPC scheme.

Theorem 1 (Stability of MPC scheme – optimal case). *Suppose that Assumptions 1-3 are satisfied and consider the compact set*

$$\Gamma_\alpha = \{x \in \mathbb{R}^n : J^*(x) \leq \alpha\} \,, \quad \alpha := \beta\Big(1 + \frac{m_l}{M_V} T\Big) \,. \tag{10}$$

Then, for all $x_0 \in \Gamma_\alpha$ *the following holds:*

1. *For all MPC steps, it holds that* $x_k \in \Gamma_\alpha$. *Moreover, the endpoint of the optimal state trajectory* $\bar{x}_k^*(\tau)$, $\tau \in [0, T]$ *reaches the CLF region, i.e.* $\bar{x}_k^*(T) \in S_\beta$.
2. Γ_α *contains the CLF region, i.e.* $S_\beta \subseteq \Gamma_\alpha$.
3. *The optimal cost satisfies*

$$J^*(\bar{x}_k^*(\Delta t)) \leq J^*(x_k) - \int_0^{\Delta t} l(\bar{x}_k^*(\tau), \bar{u}_k^*(\tau))\, \mathrm{d}\tau \quad \forall\, x_k \in \Gamma_\alpha \,. \tag{11}$$

4. *The origin of the system* (1) *under the optimal MPC law* (7) *is asymptotically stable in the sense that the closed-loop trajectories* (8) *satisfy* $\lim_{t\to\infty} ||x(t)|| = 0$.

The single statements 1-4 in Theorem 1 are discussed in the following:

1. The sublevel set Γ_α defines the domain of attraction for the MPC scheme without terminal constraints (Graichen & Kugi, 2010; Limon et al., 2006). The proof of this statement is given in Appendix A.
2. Although α in (10) leads to a rather conservative estimate of Γ_α due to the nature of the proof (see Appendix A), it nevertheless reveals that Γ_α can be enlarged by increasing the horizon length T.

3. The decrease condition (11) for the optimal cost at the next point $x_{k+1} = \bar{x}_k^*(\Delta t)$ follows from the CLF property (9) on the set S_β (Jadbabaie et al., 2001). Indeed, consider the trajectories

$$\hat{x}(\tau) = \begin{cases} \bar{x}_k^*(\tau + \Delta t), & \tau \in [0, T-\Delta t) \\ \bar{x}^q(\tau - T + \Delta t), & \tau \in [T-\Delta t, T] \end{cases}, \quad \hat{u}(\tau) = \begin{cases} \bar{u}_k^*(\tau + \Delta t), & \tau \in [0, T-\Delta t) \\ \bar{u}^q(\tau - T + \Delta t), & \tau \in [T-\Delta t, T] \end{cases}$$

where $\bar{x}^q(\tau)$ with $\bar{x}^q(0) = \bar{x}_k^*(T)$ is the state trajectory that results from applying the local CLF law $\bar{u}^q(\tau) = q(\bar{x}^q(\tau))$. Note that $\bar{x}^q(\tau) \in S_\beta$ for all $\tau \geq 0$, i.e. S_β ist positive invariant due to the definition of S_β and the CLF inequality (9) that can be expressed in the form

$$\frac{\mathrm{d}}{\mathrm{d}\tau} V(\bar{x}^q(\tau)) \leq -l(\bar{x}^q(\tau), \bar{u}^q(\tau)) \,. \tag{12}$$

Hence, the following estimates hold

$$\begin{aligned} J^*(x_k^*(\Delta t)) &\leq \int_0^T l(\hat{x}(\tau), \hat{u}(\tau)) \, \mathrm{d}\tau + V(\hat{x}(T)) \\ &= J^*(x_k) - \int_0^{\Delta t} l(\bar{x}_k^*(\tau), \bar{u}_k^*(\tau)) \, \mathrm{d}\tau \\ &\quad + \underbrace{V(\bar{x}^q(\Delta t)) - V(\bar{x}^q(0)) + \int_0^{\Delta t} l(\bar{x}^q(\tau), \bar{u}^q(\tau)) \, \mathrm{d}\tau}_{\leq 0} \,. \end{aligned} \tag{13}$$

4. Based on (11), Barbalat's Lemma allows one to conclude that the closed-loop trajectories (8) satisfy $\lim_{t\to\infty} ||x(t)|| = 0$, see e.g. Chen & Allgöwer (1998); Fontes (2001). Note that this property is weaker than asymptotic stability in the sense of Lyapunov, which can be proved if the optimal cost $J^*(x_k)$ is continuously differentiable (Findeisen, 2006; Fontes et al., 2007).

3. Suboptimal MPC for real-time feasibility

In practice, the exact solution of the receding horizon optimal control problem is typically approximated by a sufficiently accurate numerical solution of a suitable optimization algorithm. If the sampling time Δt is large enough, this numerical approximation will be sufficiently close to the optimal MPC case considered in the last section. However, for large-scale systems or highly dynamical systems, an accurate near-optimal solution often cannot be determined fast enough. This problem, often encountered in practice, gives rise to suboptimal MPC strategies, where an approximate solution is computed in each sampling step. This section develops the necessary changes and differences to the ideal case due to an incremental solution of the underlying OCP solution for a class of optimization algorithms.

3.1 Suboptimal solution strategy

Several suboptimal MPC strategies were already mentioned in the introduction (Cannon & Kouvaritakis, 2002; DeHaan & Guay, 2007; Diehl et al., 2005; Lee et al., 2002; Michalska & Mayne, 1993; Scokaert et al., 1999). Moreover, a suboptimal MPC scheme without terminal constraints – as considered in this chapter – was investigated in Graichen & Kugi (2010).

Instead of relying on one particular optimization method, it is assumed in Graichen & Kugi (2010) that an optimization algorithm exists that computes a control and state trajectory

$$\bar{u}_k^{(j)}(\tau) := \bar{u}^{(j)}(\tau; x_k), \quad \bar{x}_k^{(j)}(\tau) := \bar{x}^{(j)}(\tau; x_k, \bar{u}_k^{(j)}), \tau \in [0, T], \quad j \in \mathbb{N}_0^+ \tag{14}$$

in each iteration j while satisfying a linear rate of convergence

$$J(x_k, \bar{u}_k^{(j+1)}) - J^*(x_k) \leq p\left(J(x_k, \bar{u}_k^{(j)}) - J^*(x_k)\right), \quad j \in \mathbb{N}_0^+ \tag{15}$$

with a convergence rate $p \in (0, 1)$ and the limit $\lim_{j\to\infty} J(x_k, \bar{u}_k^{(j)}) = J^*(x_k)$.

In the spirit of a real-time feasible MPC implementation, the optimization algorithm is stopped after a fixed number of iterations, $j = N$, and the first part of the suboptimal control trajectory $\bar{u}_k^{(N)}(\tau)$ is used as control input

$$u(t_k + \tau) = \bar{u}_k^{(N)}(\tau), \quad \tau \in [0, \Delta t), \quad k \in \mathbb{N}_0^+ \tag{16}$$

to the system (1). In the absence of model errors and disturbances the next point x_{k+1} is given by $x_{k+1} = \bar{x}_k^{(N)}(\Delta t)$ and the closed-loop trajectories are

$$\begin{aligned} x(t) &= x(t_k + \tau) = \bar{x}^{(N)}(\tau; x_k), \\ u(t) &= u(t_k + \tau) = \bar{u}^{(N)}(\tau; x_k), \quad \tau \in [0, \Delta t), \quad k \in \mathbb{N}_0^+. \end{aligned} \tag{17}$$

Compared to the "optimal" MPC case, where the optimal trajectories (6) are computed in each MPC step k, the trajectories (14) are suboptimal, which can be characterized by the *optimization error*

$$\Delta J^{(N)}(x_k) := J(\bar{u}_k^{(N)}, x_k) - J^*(x_k) \geq 0. \tag{18}$$

In the next MPC step, the last control $\bar{u}_k^{(N)}$ (shifted by Δt) is re-used to construct a new initial control

$$\bar{u}_{k+1}^{(0)}(\tau) = \begin{cases} \bar{u}_k^{(N)}(\tau + \Delta t) & \text{if } \tau \in [0, T - \Delta t) \\ q(\bar{x}_k^{(N)}(T)) & \text{if } \tau \in [T - \Delta t, T], \end{cases} \tag{19}$$

where the last part of $\bar{u}_{k+1}^{(0)}$ is determined by the local CLF feedback law. The goal of the suboptimal MPC strategy therefore is to successively reduce the optimization error $\Delta J^{(N)}(x_k)$ in order to improve the MPC scheme in terms of optimality. Figure 1 illustrates this context.

3.2 Stability and incremental improvement

Several further assumptions are necessary to investigate the stability and the evolution of the optimization error for the suboptimal MPC scheme.

Assumption 4. *The optimal control law in* (7) *is locally Lipschitz continuous.*

Assumption 5. *For every $\bar{u} \in \mathcal{U}_{[0,T]}$, the cost $J(x_k, \bar{u})$ is twice continuously differentiable in x_k.*

Assumption 6. *For all $\bar{u} \in \mathcal{U}_{[0,T]}$ and all $x_k \in \Gamma_\alpha$, the cost $J(x_k, \bar{u})$ satisfies the quadratic growth condition $C||\bar{u} - \bar{u}_k^*||^2_{L_2^m[0,T]} \leq J(x_k, \bar{u}) - J^*(x_k)$ for some constant $C > 0$.*

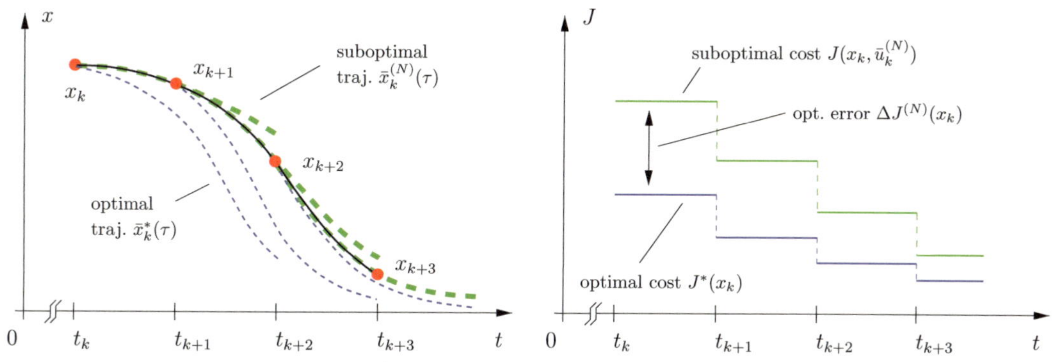

Fig. 1. Illustration of the suboptimal MPC implementation.

Assumption 6 is always satisfied for linear systems with quadratic cost functional as proved in Appendix B. In general, the quadratic growth property in Assumption 6 represents a smoothness assumption which, however, is weaker than assuming strong convexity (it is well known that strong convexity on a compact set implies quadratic growth, see, e.g., Allaire (2007) and Appendix B). [2]

The stability analysis for the suboptimal MPC case is more involved than in the "optimal" MPC case due to the non-vanishing optimization error $\Delta J^{(N)}(x_k)$. An important question in this context is under which conditions the CLF region S_β can be reached by the suboptimal state trajectory $\bar{x}_k^{(N)}(\tau)$. The following theorem addresses this question and also gives sufficient conditions for the stability of the suboptimal MPC scheme.

Theorem 2 (Stability of MPC scheme – suboptimal case). *Suppose that Assumptions 1-6 are satisfied and consider the subset of the domain* (10)

$$\Gamma_{\hat{\alpha}} = \{x \in \mathbb{R}^n : J^*(x) \leq \hat{\alpha}\}, \quad \hat{\alpha} = \frac{m_V}{4M_V}\alpha < \alpha. \tag{20}$$

Then, there exists a minimum number of iterations $\hat{N} \geq 1$ and a maximum admissible optimization error $\Delta\hat{J} \geq 0$, such that for all $x_0 \in \Gamma_{\hat{\alpha}}$ and all initial control trajectories $\bar{u}_0^{(0)} \in \mathcal{U}_{[0,T]}$ satisfying $\Delta J^{(0)}(x_0) \leq p^{-N}\Delta\hat{J}$ the following holds:

1. *For all MPC steps, it holds that $x_k \in \Gamma_{\hat{\alpha}}$. Moreover, the endpoint of the (suboptimal) state trajectory $\bar{x}_k^{(N)}(\tau)$, $\tau \in [0,T]$ reaches the CLF region, i.e. $\bar{x}_k^{(N)}(T) \in S_\beta$.*
2. *$\Gamma_{\hat{\alpha}}$ contains the CLF region, i.e. $S_\beta \subseteq \Gamma_{\hat{\alpha}}$, if the horizon length satisfies $T \geq (\frac{4M_V}{m_V} - 1)\frac{M_V}{m_l}$.*
3. *The origin of the system* (1) *under the suboptimal MPC law* (16) *is exponentially stable.*
4. *The optimization error* (18) *decays exponentially.*

The proof of Theorem 2 consists of several intermediate lemmas and steps that are given in details in Graichen & Kugi (2010). The statements 1-4 in Theorem 2 summarize several important points that deserve some comments.

[2] A simple example is the function $f(x) = x^2 + 10\sin^2 x$ with the global minimum $f(x^*) = 0$ at $x^* = 0$. Let x be restricted to the interval $x \in [-5,5]$. Clearly, the quadratic growth property $\frac{1}{2}|x - x^*|^2 \leq f(x) - f(x^*)$ is satisfied for $x \in [-5,5]$ although $f(x)$ is not convex on this interval.

1. The reduced size of $\Gamma_{\hat{\alpha}}$ compared to Γ_{α} is the necessary "safety" margin to account for the suboptimality of the trajectories (6) characterized by $\Delta J^{(N)}(x_k)$. Thus, the domain of attraction $\Gamma_{\hat{\alpha}}$ together with an admissible upper bound on the optimization error guarantees the reachability of the CLF region S_{β}.
2. An interesting fact is that it can still be guaranteed that $\Gamma_{\hat{\alpha}}$ is at least as large as the CLF region S_{β} provided that the horizon time T satisfies a lower bound that depends on the quadratic estimates (5) of the integral and terminal cost functions. It is apparent from the bound $T \geq (\frac{4M_V}{m_V} - 1)\frac{M_V}{m_l}$ that the more dominant the terminal cost $V(x)$ is with respect to the integral cost function $l(x,u)$, the larger this bound on the horizon length T will be.
3. The minimum number of iterations $\hat{N}$ for which stability can be guaranteed ensures – roughly speaking – that the numerical speed of convergence is faster than the system dynamics. In the proof of the theorem (Graichen & Kugi, 2010), the existence of the lower bound $\hat{N}$ is shown by means of Lipschitz estimates, which usually are too conservative to be used for design purposes. For many practical problems, however, one or two iterations per MPC step are sufficient to ensure stability and a good control performance.
4. The exponential reduction of the optimization error $\Delta J^{(N)}(x_k)$ follows as part of the proof of stability and reveals the incremental improvement of the suboptimal MPC scheme over the MPC runtime.

4. Gradient projection method

The efficient numerical implementation of the MPC scheme is of importance to guarantee the real-time feasibility for fast dynamical systems. This section describes the well-known gradient projection in optimal control as well as its suboptimal implementation in the context of MPC.

4.1 Optimality conditions and algorithm

The MPC formulation without terminal constraints has particular advantages for the structure of the optimality conditions of the OCP (3). To this end, we define the Hamiltonian

$$H(x,\lambda,u) = l(x,u) + \lambda^{\mathsf{T}} f(x,u) \tag{21}$$

with the adjoint state $\lambda \in \mathbb{R}^n$. Pontryagin's Maximum Principle[3] states that if $\bar{u}_k^*(\tau), \tau \in [0,T]$ is an optimal control for OCP (3), then there exists an adjoint trajectory $\bar{\lambda}_k^*(\tau), \tau \in [0,T]$ such that $\bar{x}_k^*(\tau)$ und $\bar{\lambda}_k^*(\tau)$ satisfy the canonical boundary value problem (BVP)

$$\dot{\bar{x}}_k^*(\tau) = f(\bar{x}_k^*(\tau), \bar{u}_k^*(\tau)), \qquad \bar{x}_k^*(0) = x_k \tag{22}$$

$$\dot{\bar{\lambda}}_k^*(\tau) = -H_x(\bar{x}_k^*(\tau), \bar{\lambda}_k^*(\tau), \bar{u}_k^*(\tau)), \quad \bar{\lambda}_k^*(T) = V_x(\bar{x}_k^*(T)) \tag{23}$$

and $\bar{u}_k^*(\tau)$ minimizes the Hamiltonian for all times $\tau \in [0,T]$, i.e.

$$H(\bar{x}_k^*(\tau), \bar{\lambda}_k^*(\tau), \bar{u}_k^*(\tau)) \leq H(\bar{x}_k^*(\tau), \bar{\lambda}_k^*(\tau), u), \quad \forall\, u \in [u^-, u^+], \quad \forall\, \tau \in [0,T]. \tag{24}$$

[3] The general formulation of Pontryagin's Maximum Principle often uses the Hamiltonian definition $H(x,\lambda,u,\lambda_0) = \lambda_0 l(x,u) + \lambda^{\mathsf{T}} f(x,u)$, where λ_0 accounts for "abnormal" problems as, for instance, detailed in Hsu & Meyer (1968). Typically, λ_0 is set to $\lambda_0 = 1$, which corresponds to the definition (21).

The functions H_x and V_x denote the partial derivatives of H and V with respect to x. The minimization condition (24) also allows one to conclude that the partial derivative $H_u = [H_{u,1}, \ldots, H_{u,m}]^\mathsf{T}$ of the Hamiltonian with respect to the control $u = [u_1, \ldots, u_m]^\mathsf{T}$ has to satisfy

$$H_{u,i}(\bar{x}_k^*(\tau), \bar{\lambda}_k^*(\tau), \bar{u}_k^*(\tau)) \begin{cases} > 0 & \text{if } \bar{u}_{k,i}^*(\tau) = u_i^- \\ = 0 & \text{if } \bar{u}_{k,i}^*(\tau) \in (u_i^-, u_i^+), \quad i = 1, \ldots, m, \quad \tau \in [0, T]. \\ < 0 & \text{if } \bar{u}_{k,i}^*(\tau) = u_i^+ \end{cases}$$

The adjoint dynamics in (23) possess n terminal conditions which is due to the free endpoint formulation of OCP (3). This property is taken advantage of by the gradient method, which solves the canonical BVP (22)-(23) iteratively forward and backward in time. Table 1 summarizes the algorithm of the gradient (projection) method.

The search direction $\bar{s}_k^{(j)}(\tau)$, $\tau \in [0, T]$ is the direction of improvement for the current control $\bar{u}_k^{(j)}(\tau)$. The step size $\alpha_k^{(j)}$ is computed in the subsequent line search problem (28) in order to achieve the maximum possible descent of the cost functional (3a). The function

---

1) **Initialization for $j = 0$:**

- Set convergence tolerance ε_J (e.g. $\varepsilon_J = 10^{-6}$)
- Choose initial control trajectory $\bar{u}_k^{(0)} \in \mathcal{U}_{[0,T]}$
- Integrate forward in time

$$\dot{\bar{x}}_k^{(0)}(\tau) = f(\bar{x}_k^{(0)}(\tau), \bar{u}_k^{(0)}(\tau)), \quad \bar{x}_k^{(0)}(0) = x_k \tag{25}$$

2) **Gradient step: While $j \leq N$ Do**

- Integrate backward in time

$$\dot{\bar{\lambda}}_k^{(j)}(\tau) = -H_x(\bar{x}_k^{(j)}(\tau), \bar{\lambda}_k^{(j)}(\tau), \bar{u}_k^{(j)}(\tau)), \quad \bar{\lambda}_k^{(j)}(T) = V_x(\bar{x}_k^{(j)}(T)) \tag{26}$$

- Compute the search direction

$$\bar{s}_k^{(j)}(\tau) = -H_u(\bar{x}_k^{(j)}(\tau), \bar{\lambda}_k^{(j)}(\tau), \bar{u}_k^{(j)}(\tau)), \quad \tau \in [0, T] \tag{27}$$

- Compute the step size $\alpha_k^{(j)}$ by (approximately) solving the line search problem

$$\alpha_k^{(j)} = \arg\min_{\alpha > 0} J\Big(x_k, \psi(\bar{u}_k^{(j)} + \alpha \bar{s}_k^{(j)})\Big) \tag{28}$$

- Compute the new control trajectory

$$\bar{u}_k^{(j+1)}(\tau) = \psi\Big(\bar{u}_k^{(j)}(\tau) + \alpha_k^{(j)} \bar{s}_k^{(j)}(\tau)\Big) \tag{29}$$

- Integrate forward in time

$$\dot{\bar{x}}_k^{(j+1)}(\tau) = f(\bar{x}_k^{(j+1)}(\tau), \bar{u}_k^{(j+1)}(\tau)), \quad \bar{x}_k^{(j+1)}(0) = x_k \tag{30}$$

- Quit if $|J(x_k, \bar{u}_k^{(j+1)}) - J(x_k, \bar{u}_k^{(j)})| \leq \varepsilon_J$. Otherwise set $j \leftarrow j + 1$ and return to 2).

Table 1. Gradient projection method for solving OCP (3).

$\psi = [\psi_1, \ldots, \psi_m]^\mathsf{T}$ in (28) represents a projection function of the form

$$\psi_i(u_i) = \begin{cases} u_i^- & \text{if } u_i < u_i^- \\ u_i^+ & \text{if } u_i > u_i^+ \\ u_i & \text{else} \end{cases}, \quad i = 1, \ldots, m \tag{31}$$

which guarantess the adherence of the input constraints $[u^-, u^+]$. For the real-time implementation within a suboptimal MPC scheme, the line search problem (28) can be solved in an approximate manner (see Section 4.2). Finally, the control trajectory $\bar{u}_k^{(j+1)}(\tau), \tau \in [0, T]$ follows from evaluating (29) with $\bar{s}_k^{(j)}(\tau)$ and the step size $\alpha_k^{(j)}$.

The convergence properties of the gradient (projection) method are investigated, for instance, in Dunn (1996); Leese (1977); Nikol'skii (2007). In particular, Dunn (1996) proved under certain convexity and regularity assumptions that the gradient method exhibits a linear rate of convergence of the form (15).

4.2 Adaptive line search

The line search (28) represents a scalar optimization problem that is often solved approximately. The most straightforward way is to use a fixed step size α throughout all gradient iterations. This, however, usually leads to a slow rate of convergence.

An attractive alternative to a constant step size is to use a polynomial approximation with an underlying interval adaptation. To this end, the cost functional $J(x_k, \psi(\bar{u}_k^{(j)} + \alpha \bar{s}_k^{(j)}))$ in the line search problem (28) is evaluated at three sample points

$$\alpha_1 < \alpha_2 < \alpha_3 \quad \text{with} \quad \alpha_2 = (\alpha_1 + \alpha_3)/2 \tag{32}$$

that are used to construct a quadratic polynomial approximation $g(\alpha)$ of the form

$$J\Big(x_k, \psi(\bar{u}_k^{(j)} + \alpha \bar{s}_k^{(j)})\Big) \approx g(\alpha) := c_0 + c_1\alpha + c_2\alpha^2 \,. \tag{33}$$

The coefficients c_0, c_1, c_2 are obtained by solving the set of equations

$$J\Big(x_k, \psi(\bar{u}_k^{(j)} + \alpha_i \bar{s}_k^{(j)})\Big) =: J_i = g(\alpha_i)\,, \quad i = 1, 2, 3 \tag{34}$$

with the explicit solution

$$\begin{aligned} c_0 &= \frac{\alpha_1(\alpha_1 - \alpha_2)\alpha_2 J_3 + \alpha_2\alpha_3(\alpha_2 - \alpha_3)J_1 + \alpha_1\alpha_3(\alpha_3 - \alpha_1)J_2}{(\alpha_1 - \alpha_2)(\alpha_1 - \alpha_3)(\alpha_2 - \alpha_3)} \\ c_1 &= \frac{(\alpha_2^2 - \alpha_1^2)J_3 + (\alpha_1^2 - \alpha_3^2)J_2 + (\alpha_3^2 - \alpha_2^2)J_1}{(\alpha_1 - \alpha_2)(\alpha_1 - \alpha_3)(\alpha_2 - \alpha_3)} \\ c_2 &= \frac{(\alpha_1 - \alpha_2)J_3 + (\alpha_2 - \alpha_3)J_1 + (\alpha_3 - \alpha_1)J_2}{(\alpha_1 - \alpha_2)(\alpha_1 - \alpha_3)(\alpha_2 - \alpha_3)}\,. \end{aligned} \tag{35}$$

If $c_2 > 0$, then the polynomial $g(\alpha)$ has a minimum at the point

$$\hat{\alpha} = -\frac{c_1}{2c_2}. \tag{36}$$

If in addition $\hat{\alpha}$ lies inside the interval $[\alpha_1, \alpha_3]$, then $\hat{\alpha} = \alpha_k^{(j)}$ approximately solves the line search problem (28). Otherwise, $\alpha_k^{(j)}$ is set to one of the interval bounds α_1 or α_3. In this case, the interval $[\alpha_1, \alpha_3]$ can be adapted by a scaling factor to track the minimum point of the line search problem (28) over the single gradient iterations. Table 2 summarizes the overall algorithm for the approximate line search and the interval adaptation.

In general, the gradient method in Table 1 is stopped if the convergence criterion is fulfilled for some tolerance $\varepsilon_J > 0$. In practice this can lead to a large number of iterations that moreover varies from one MPC iteration to the next. In order to ensure a real-time feasible MPC implementation, the gradient algorithm is stopped after N iterations and the re-initialization of the algorithm is done as outlined in Section 3.1.

---

1) **Initialization: Default values and tolerances**
- Set polynomial tolerances $\varepsilon_c, \varepsilon_g$ (e.g. $\varepsilon_c = 10^{-5}, \varepsilon_g = 10^{-6}$)
- Set initial line search interval (32) (e.g. $\alpha_1 = 10^{-2}, \alpha_3 = 10^{-1}$)
- Set interval adaptation factors κ^-, κ^+ (e.g. $\kappa^- = \frac{2}{3}, \kappa^+ = \frac{3}{2}$)
- Set interval adaptation tolerances $\varepsilon_\alpha^-, \varepsilon_\alpha^+$ (e.g. $\varepsilon_\alpha^- = 0.1, \varepsilon_\alpha^+ = 0.9$)
- Set interval adaptation limits $\alpha_{\min}, \alpha_{\max}$ (e.g. $\alpha_{\min} = 10^{-5}, \alpha_{\max} = 1.0$)

2) **Approximate line search**
- Compute the cost values $J_i := J(x_k, \psi(u_k^{(j)} + \alpha_i s_k^{(j)}))$ at the sample points (32)
- Compute the polynomial coefficients (35) and the candidate point (36)
- Compute the approximate step size $\alpha_k^{(j)}$ according to

$$\text{if } c_2 > \varepsilon_c: \quad \alpha_k^{(j)} = \begin{cases} \alpha_1 & \text{if } \hat{\alpha} < \alpha_1 \\ \alpha_3 & \text{if } \hat{\alpha} > \alpha_3 \\ \hat{\alpha} & \text{else} \end{cases} \tag{37}$$

$$\text{else } (c_2 \le \varepsilon_c): \quad \alpha_k^{(j)} = \begin{cases} \alpha_1 & \text{if } J_1 + \varepsilon_g \le \min\{J_2, J_3\} \\ \alpha_3 & \text{if } J_3 + \varepsilon_g \le \min\{J_1, J_2\} \\ \alpha_2 & \text{else} \end{cases} \tag{38}$$

- Adapt the line search interval $[\alpha_1, \alpha_3]$ for the next gradient iteration according to

$$[\alpha_1, \alpha_3] \leftarrow \begin{cases} \kappa^+[\alpha_1, \alpha_3] & \text{if } \hat{\alpha} \ge \alpha_1 + \varepsilon_\alpha^+(\alpha_3 - \alpha_1) \text{ and } \alpha_3 \le \alpha_{\max} \\ \kappa^-[\alpha_1, \alpha_3] & \text{if } \hat{\alpha} \le \alpha_1 + \varepsilon_\alpha^-(\alpha_3 - \alpha_1) \text{ and } \alpha_1 \ge \alpha_{\min} \\ [\alpha_1, \alpha_3] & \text{else} \end{cases}, \quad \alpha_2 \leftarrow \frac{\alpha_1 + \alpha_3}{2} \tag{39}$$

Table 2. Adaptive line search for the gradient algorithm in Table 1.

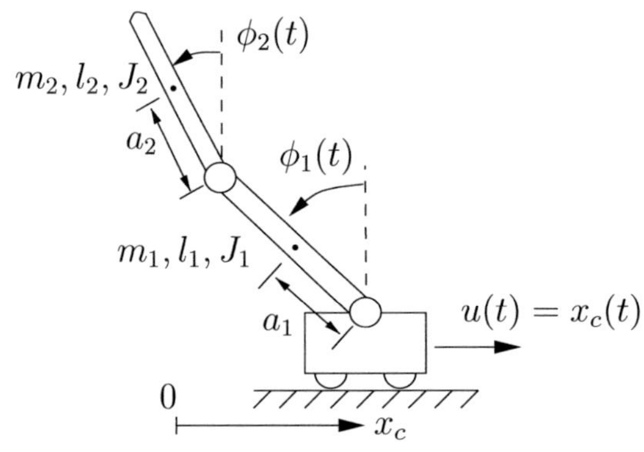

Fig. 2. Inverted double pendulum on a cart.

5. Example – Inverted double pendulum

The inverted double pendulum on a cart is a benchmark problem in control theory due to its highly nonlinear and nonminimum-phase dynamics and its instability in the upward (inverted) position. The double pendulum in Figure 2 consists of two links with the lengths l_i and the angles ϕ_i, $i = 1,2$ to the vertical direction. The displacement of the cart is given by x_c. The mechanical parameters are listed in Table 3 together with their corresponding values (Graichen et al., 2007). The double pendulum is used in this section as benchmark example for the suboptimal MPC scheme and the gradient algorithm in order to show its performance for a real-time MPC implementation.

5.1 Equations of motion and MPC formulation

Applying the Lagrangian formalism to the double pendulum leads to the equations of motion (Graichen et al., 2007)

$$M(\phi)\ddot{\phi} + c(\phi, \dot{\phi}, \ddot{x}_c) = 0 \tag{40}$$

with the generalized coordinates $\phi = [\phi_1, \phi_2]^\mathsf{T}$ and the functions

$$M(\phi) = \begin{bmatrix} J_1 + a_1^2 m_1 + l_1^2 m_2 & a_2 l_1 m_2 \cos(\phi_1 - \phi_2) \\ a_2 l_1 m_2 \cos(\phi_1 - \phi_2) & J_2 + a_2^2 m_2 \end{bmatrix} \tag{41a}$$

$$c(\phi, \dot{\phi}) = \begin{bmatrix} d_1\dot{\phi}_1 + d_2(\dot{\phi}_1 - \dot{\phi}_2) + l_1 m_2 a_2 \sin(\phi_1 - \phi_2)\dot{\phi}_2^2 - (a_1 m_1 + l_1 m_2)\left[g \sin(\phi_1) + \cos(\phi_2)\ddot{x}_c\right] \\ d_2(\dot{\phi}_2 - \dot{\phi}_1) - a_2 m_2 \left[g \sin(\phi_2) + l_1 \sin(\phi_1 - \phi_2)\dot{\phi}_1^2 + \cos(\phi_2)\ddot{x}_c\right] \end{bmatrix}. \tag{41b}$$

The acceleration of the cart $\ddot{x}_c$ serves as control input u. Thus, the overall model of the double pendulum can be written as the second-order ordinary differential equations (ODE)

$$\begin{aligned} \ddot{x}_c &= u \\ \ddot{\phi} &= -M^{-1}(\phi)c(\phi, \dot{\phi}, u). \end{aligned} \tag{42}$$

The acceleration of the cart is limited by the constraints

$$u \in [-6, +6] \text{ m/s}^2 . \tag{43}$$

Pendulum link	inner $i=1$	outer $i=2$
length l_i [m]	0.323	0.480
distance to center of gravity a_i [m]	0.215	0.223
mass m_i [kg]	0.853	0.510
moment of inertia J_i [N m s^2]	0.013	0.019
friction constant d_i [N m s]	0.005	0.005

Table 3. Mechanical parameters of the double pendulum in Figure 2.

With the state vector $x = [x_c, \dot{x}_c, \phi_1, \dot{\phi}_1, \phi_2, \dot{\phi}_2]^\mathsf{T}$, the second-order ODEs (42) can be written as the general nonlinear system

$$\dot{x} = f(x, u), \quad x(0) = x_0 . \tag{44}$$

For the MPC formulation, a quadratic cost functional (3a)

$$J(x_k, \bar{u}) = \Delta\bar{x}^\mathsf{T}(T) P \Delta\bar{x}(T) + \int_0^T \Delta\bar{x}^\mathsf{T}(\tau) Q \Delta\bar{x}(\tau) + \Delta\bar{u}^\mathsf{T}(\tau) R \Delta\bar{u}(\tau)\, \mathrm{d}\tau , \tag{45}$$

with $\Delta\bar{x} = \bar{x} - x_{SP}$ and $\Delta\bar{u} = \bar{u} - u_{SP}$ is used, which penalizes the distance to a desired setpoint (x_{SP}, u_{SP}), i.e. $0 = f(x_{SP}, u_{SP})$. The symmetric and positive definite weighting matrices Q, R in the integral part of (45) are chosen as

$$Q = \mathrm{diag}(10, 0.1, 1, 0.1, 1, 0.1) , \quad R = 0.001 . \tag{46}$$

The CLF condition in Assumption 3 is approximately satisfied by solving the Riccati equation

$$PA + A^\mathsf{T} P - P b R^{-1} b^\mathsf{T} P + Q = 0 , \tag{47}$$

where $A = \frac{\partial f}{\partial x}\big|_{x_{SP}, u_{SP}}$ and $b = \frac{\partial f}{\partial u}\big|_{x_{SP}, u_{SP}}$ describe the linearization of the system (44) around the setpoint (x_{SP}, u_{SP}).[4] The sampling time Δt and the prediction horizon T are set to

$$\Delta t = 1 \text{ ms} , \quad T = 0.3 \text{ s} \tag{48}$$

to account for the fast dynamics of the double pendulum and the highly unstable behavior in the inverted position.

5.2 Simulation results

The suboptimal MPC scheme together with the gradient method were implemented as Cmex functions under MATLAB. The functions that are required in the gradient method are

[4] For the linearized (stabilizable) system $\Delta\dot{x} = A\Delta x + b\Delta u$, it can be shown that the CLF inequality (9) is exactly fulfilled (in fact, (9) turns into an equality) for the terminal cost $V(x) = \Delta x^\mathsf{T} P \Delta x$ and the linear (unconstrained) feedback law $q(x) = -R^{-1} b^\mathsf{T} P \Delta x$ with P following from the Riccati equation (47).

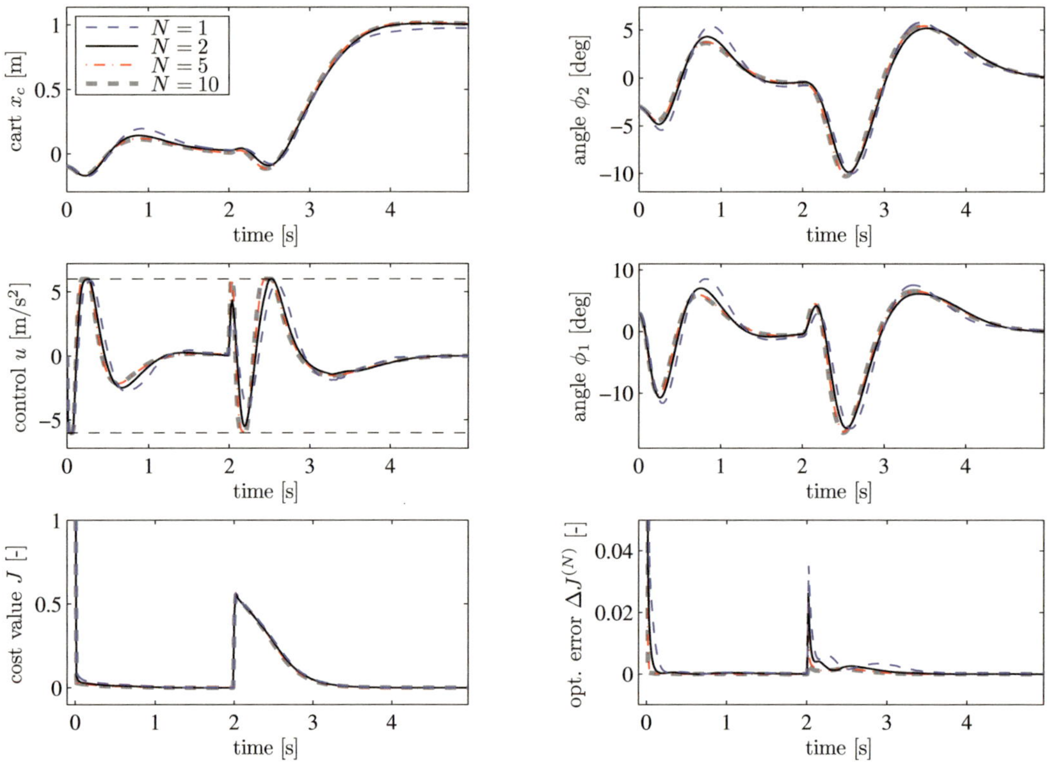

Fig. 3. MPC results for the double pendulum on a cart.

computed under the computer algebra system MATHEMATICA and are exported to MATLAB as optimized C code. The numerical integrations of the canonical equations (25)-(30) are performed by discretizing the time interval $[0, T]$ with a fixed number of 30 equidistant points and using a second order Runge-Kutta method. The nonlinear model (44), respectively (41), is used within the MPC scheme as well as for the simulation of the pendulum.

The considered simulation scenario consists of an initial error around the origin ($x_{SP} = 0$, $u_{SP} = 0$) and a subsequent setpoint step of 1 m in the cart position at time $t = 2$ s ($x_{SP} = [1\text{ m}, 0, 0, 0, 0, 0]^\mathsf{T}$, $u_{SP} = 0$). Figure 3 shows the simulation results for a two-stage scenario (initial error and setpoint change at $t = 2$ s). Already the case of one gradient iteration per sampling step ($N = 1$) leads to a good control performance and a robust stabilization of the double pendulum. Increasing N results in a more aggressive control behavior and a better exploitation of the control constraints (43).

The lower plots in Figure 3 show the (discrete-time) profiles of the cost value $J(x_k, \bar{u}_k^{(N)})$ and of the optimization error $\Delta J^{(N)}(x_k) = J(x_k, \bar{u}_k^{(N)}) - J^*(x_k)$. In order to determine $\Delta J^{(N)}(x_k)$, the optimal cost $J^*(x_k)$ was computed in each step x_k by solving the OCP (3) for the double pendulum with a collocation-based optimization software. It is apparent from the respective plots in Figure 3 that the cost as well as the optimization error rapidly converge to zero which illustrates the exponential stability of the double pendulum in closed-loop and the incremental improvement of the algorithm. It is also seen in these plots that the performance improvement

MPC iter. N / sampling step	CPU time [ms] / sampling step	mean cost value [–]
1	0.053	0.0709
2	0.095	0.0641
3	0.133	0.0632
5	0.212	0.0610
10	0.405	0.0590

Table 4. CPU time consumption of the real-time MPC scheme for different numbers of gradient iterations N per sampling step.

between $N = 1$ and $N = 10$ iterations per sampling step are comparatively small compared to the increase of numerical load.

To investigate this point more precisely, Table 4 lists the required CPU time for different MPC settings. The computations were performed on a computer with an Intel i7 CPU (M620, 2.67 GHz) [5], 4 GB of memory, and the operating system MS Windows 7 (64 bit). The overall MPC scheme compiled as Cmex function under MATLAB 2010b (64 bit). All evaluated tests in Table 4 show that the required CPU times are well below the actual sampling time of $\Delta t = 1\,\text{ms}$. The CPU times are particularly remarkable in view of the high complexity of the nonlinear pendulum model (40)-(42), which illustrates the real-time feasibility of the suboptimal MPC scheme. The last column in Table 4 shows the average cost value that is obtained by integrating the cost profiles in Figure 3 and dividing through the simulation time of 5 s. This index indicates that the MPC scheme increases in terms of control performance for larger numbers of N.

From these numbers and the simulation profiles in Figure 3, the conclusion can be drawn that $N = 2$ gradient iterations per MPC step represents a good compromise between control performance and the low computational demand of approximately 100 μs per MPC step.

6. Conclusions

Suboptimal solution strategies are efficient means to reduce the computational load for a real-time MPC implementation. The suboptimal solution from the previous MPC step is used for a warm-start of the optimization algorithm in the next run with the objective to reduce the suboptimality over the single MPC steps. Section 3 provides theoretical justifications for a suboptimal MPC scheme with a fixed number of iterations per sampling step.

A suitable optimization algorithm is the gradient method in optimal control, which allows for a time and memory efficient calculation of the single MPC iterations and makes the overall MPC scheme suitable for very fast or high dimensional dynamical systems. The control performance and computational efficiency of the gradient method is illustrated in Section 5 for a highly nonlinear and complex model of a double pendulum on a cart. The suboptimal MPC scheme based on a real-time implementation of the gradient method was

[5] Only one core of the i7 CPU was used for the computations.

also experimentally validated for a laboratory crane (Graichen et al., 2010) and for a helicopter with three degrees-of-freedom (Graichen et al., 2009), both experiments with sampling times of 1-2 milliseconds. The applicability of the gradient-based MPC scheme to high dimensional systems is demonstrated in (Steinböck et al., 2011) for a reheating furnace in steel industry.

7. Appendix A – Reachability of CLF region (Theorem 1)

This appendix proves the statements 1 and 2 in Theorem 1 concerning the reachability of the CLF region S_β by the MPC formulation (3) without terminal constraints. The discrete-time case was investigated in Limon et al. (2006). The following two lemmas generalize these results to continuous-time systems as considered in this chapter. Lemma 1 represents an intermediate statement that is required to derive the actual result in Lemma 2.

Lemma 1. *Suppose that Assumptions 1-3 are satisfied. If* $\bar{x}_k^*(T) \notin S_\beta$ *for any* $x_k \in \mathbb{R}^n$*, then* $\bar{x}_k^*(\tau) \notin S_\beta$ *for all times* $\tau \in [0, T]$.

Proof. The proof is accomplished by contradiction. Assume that $\bar{x}_k^*(T) \notin S_\beta$ and that there exists a time $\hat{\tau} \in [0, T)$ such that $\bar{x}_k^*(\hat{\tau}) \in S_\beta$. Starting at this point $\bar{x}_k^*(\hat{\tau})$, consider the residual problem

$$\hat{J}^*(\bar{x}_k^*(\hat{\tau})) = \min_{\bar{u} \in \mathcal{U}_{[T-\hat{\tau}]}} \left\{ V\Big(\bar{x}(T-\hat{\tau}; \bar{x}_k^*(\hat{\tau}), \bar{u})\Big) + \int_0^{T-\hat{\tau}} l\Big(\bar{x}(\tau; \bar{x}_k^*(\hat{\tau}), \bar{u}), \bar{u}(\tau)\Big)\, \mathrm{d}\tau \right\}$$

subject to the dynamics (3b), for which the optimal trajectories are $\hat{u}^*(\tau) = \bar{u}_k^*(\hat{\tau} + \tau)$ and $\hat{x}^*(\tau) = \bar{x}_k^*(\hat{\tau} + \tau)$, $\tau \in [0, T - \hat{\tau}]$ by the principle of optimality. Since $\bar{x}_k^*(\hat{\tau}) \in S_\beta$ by assumption, the CLF inequality (9) with $\bar{x}^q(0) = \bar{x}_k^*(\hat{\tau})$ leads to the lower bound

$$\begin{aligned} V(\bar{x}_k^*(\hat{\tau})) &\geq V(\bar{x}^q(T-\hat{\tau})) + \int_0^{T-\hat{\tau}} l(\bar{x}^q(\tau), \bar{u}^q(\tau))\mathrm{d}\tau \\ &\geq \hat{J}^*(\bar{x}_k^*(\hat{\tau})) \\ &\geq V(\hat{x}^*(T-\hat{\tau})) = V(\bar{x}_k^*(T)) > \beta\,. \end{aligned}$$

The last line, however, implies that $\bar{x}_k^*(\hat{\tau}) \notin S_\beta$, which contradicts the previous assumption and thus proves the lemma. □

Lemma 2. *Suppose that Assumptions 1-3 are satisfied and consider the compact set* Γ_α *defined by* (10). *Then, for all* $x_k \in \Gamma_\alpha$*, the endpoint of the optimal state trajectory satisfies* $\bar{x}_k^*(T) \in S_\beta$*. Moreover,* $S_\beta \subseteq \Gamma_\alpha$.

Proof. We will again prove the lemma by contradiction. Assume that there exists a $x_k \in \Gamma_\alpha$ such that $\bar{x}_k^*(T) \notin S_\beta$, i.e. $V(\bar{x}_k^*(T)) > \beta$. Then, Lemma 1 states that $\bar{x}_k^*(\tau) \notin S_\beta$ for all $\tau \in [0, T]$, or using (5),

$$||\bar{x}_k^*(\tau)||^2 > \frac{\beta}{M_V} \quad \forall \tau \in [0, T]\,. \tag{49}$$

This allows one to derive a lower bound on the optimal cost

$$\begin{aligned} J^*(x_k) &= V(\bar{x}_k^*(T)) + \int_0^T l(\bar{x}_k^*(\tau), \bar{u}_k^*(t))\,\mathrm{d}\tau \\ &\geq \beta + \int_0^T m_l \frac{\beta}{M_V}\mathrm{d}\tau \\ &= \beta\left(1 + \frac{m_l}{M_V}T\right) \\ &= \alpha\,. \end{aligned} \tag{50}$$

From this last line it can be concluded that $x_k \notin \Gamma_\alpha$ for all $\bar{x}_k^*(T) \notin S_\beta$. This, however, is a contradiction to the previous assumption and implies that $\bar{x}_k^*(T) \in S_\beta$ for all $x_k \in \Gamma_\alpha$. To prove that Γ_α contains the CLF region S_β, consider $x_k \in S_\beta$ and the bound on the optimal cost

$$J^*(x_k) \leq V(\bar{x}^q(T)) + \int_0^T l(\bar{x}^q(\tau), \bar{u}^q(\tau))\,\mathrm{d}t \tag{51}$$

with the CLF trajectories $\bar{x}^q(\tau)$, $\bar{x}^q(0) = x_k$, and $\bar{u}^q(\tau) = q(\bar{x}^q(\tau))$. Similar to the proof of Lemma 1, the CLF inequality (9) implies that

$$V(\bar{x}^q(T)) \leq V(\bar{x}^q(0)) - \int_0^T l(\bar{x}^q(\tau), \bar{u}^q(\tau))\mathrm{d}\tau\,. \tag{52}$$

Hence, (51)-(52) and definition (10) show that $J^*(x_k) \leq V(x_k) \leq \beta < \alpha$ for all $x_k \in S_\beta$, which proves that Γ_α contains S_β. □

8. Appendix B – Verification of Assumption 6 for linear-quadratic OCPs

The following lines show that Assumption 6 is fulfilled for OCPs of the form

$$\min_{u\in\mathcal{U}_{[0,T]}} \quad J(u) = x^\mathsf{T}(T)Px(T) + \int_0^\mathsf{T} x^\mathsf{T}(t)Qx(t) + u^\mathsf{T}(t)Ru(t)\,\mathrm{d}t\,, \tag{53}$$

$$\text{subj. to} \quad \dot{x} = Ax + Bu \quad x(0) = x_0\,, \quad x \in \mathbb{R}^n, \quad u \in \mathbb{R}^m \tag{54}$$

with the quadratic cost functional (53), the linear dynamics (54) and some initial state $x_0 \in \mathbb{R}^n$. The admissible input set $\mathcal{U}_{[0,T]}$ is assumed to be convex and the weighting matrices P, Q, R are symmetric and positiv definite. A useful property of the linear-quadratic problem is the strong convexity property (Allaire, 2007)

$$C||u - v||^2_{L_2^m[0,T]} \leq J(u) + J(v) - 2J(\tfrac{1}{2}u + \tfrac{1}{2}v) \tag{55}$$

for some constant $C > 0$ and all control functions $u, v \in \mathcal{U}_{[0,T]}$. To show this, first consider the control term of the cost functional (53) and the right-hand side of (55), which can be written in the form

$$\int_0^\mathsf{T} u^\mathsf{T}Ru + v^\mathsf{T}Rv - \frac{1}{2}(u+v)^\mathsf{T}R(u+v)\,\mathrm{d}t = \frac{1}{2}\int_0^\mathsf{T}(u-v)^\mathsf{T}R(u-v)\,\mathrm{d}t\,.$$

The same simplifications can be used for the state-dependent terms in (53) since the linear dynamics (54) ensures that the superposition of two input signals $w(t) = \frac{1}{2}u(t) + \frac{1}{2}v(t)$ yield a corresponding superposed state response $x_w(t) = \frac{1}{2}x_u(t) + \frac{1}{2}x_v(t)$ with $x_w(0) = x_0$. Hence, the right-hand side of (55) can be written as

$$\begin{aligned} J(u) + J(v) - 2J(\tfrac{1}{2}u + \tfrac{1}{2}v) &= \frac{1}{2}\Delta x^{\mathsf{T}}(T)P\Delta x(T) + \frac{1}{2}\int_0^{\mathsf{T}} \Delta x^{\mathsf{T}}(t)Q\Delta x(t)\,\mathrm{d}t \\ &\quad + \frac{1}{2}\int_0^{\mathsf{T}} \Big(u(t) - v(t)\Big)^{\mathsf{T}} R\Big(u(t) - v(t)\Big)\,\mathrm{d}t \\ &\geq C||u - v||^2_{L_2^m[0,T]} \end{aligned}$$

with $\Delta x(t) = x_u(t) - x_v(t)$ and the constant $C = \lambda_{\min}(R)/2$. Since $J(u)$ is strongly (and therefore also strictly) convex on the convex set $\mathcal{U}_{[0,T]}$, it follows from standard arguments (Allaire, 2007) that there exists a global and unique minimum point $u^* \in \mathcal{U}_{[0,T]}$. Moreover, since $\mathcal{U}_{[0,T]}$ is convex, $\frac{1}{2}(u + u^*) \in \mathcal{U}_{[0,T]}$ for all $u \in \mathcal{U}_{[0,T]}$ such that $J(\frac{1}{2}u + \frac{1}{2}u^*) \geq J(u^*)$. Hence, the strong convexity inequality (55) can be turned into the quadratic growth property

$$C||u - u^*||^2_{L_2^m[0,T]} \leq J(u) + J(u^*) - 2J(\tfrac{1}{2}u + \tfrac{1}{2}u^*) \leq J(u) - J(u^*) \quad \forall\, u \in \mathcal{U}_{[0,T]}\,.$$

This shows that Assumption 6 is indeed satisfied for linear-quadratic OCPs of the form (53).

9. Acknowledgements

This work was supported by the Austrian Science Fund under project no. P21253-N22.

10. References

Allaire, G. (2007). *Numerical Analysis and Optimization*, Oxford University Press, New York.

Berkovitz, L. (1974). *Optimal Control Theory*, Springer, New York.

Cannon, M. & Kouvaritakis, B. (2002). Efficient constrained model predictive control with asymptotic optimality, *SIAM Journal on Control and Optimization* 41(1): 60–82.

Chen, H. & Allgöwer, F. (1998). A quasi-infinite horizon nonlinear model predictive control scheme with guaranteed stability, *Automatica* 34(10): 1205–1217.

DeHaan, D. & Guay, M. (2007). A real-time framework for model-predictive control of continuous-time nonlinear systems, *IEEE Transactions on Automatic Control* 52(11): 2047–2057.

Diehl, M., Ferreau, H. & Haverbeke, N. (2009). Efficient numerical methods for nonlinear MPC and moving horizon estimation, *in* L. Magni, D. Raimondo & F. Allgöwer (eds), *Nonlinear Model Predictive Control – Towards New Challenging Applications*, pp. 391–417.

Diehl, M., Findeisen, R., Allgöwer, F., Bock, H. & Schlöder, J. (2005). Nominal stability of real-time iteration scheme for nonlinear model predictive control, *IEE Proceedings Control Theory and Applications* 152(3): 296–308.

Dunn, J. (1996). On l^2 conditions amd the gradient projection method for optimal control problems, *SIAM Journal on Control and Optimization* 34(4): 1270–1290.

Findeisen, R. (2006). *Nonlinear Model Predictive Control: A Sampled-Data Feedback Perspective*, Vol. 1087, Fortschritt-Berichte VDI Reihe 8.

Fontes, F. (2001). A general framework to design stabilizing nonlinear model predictive controllers, *Systems & Control Letters* 42(2): 127–143.

Fontes, F., Magni, L. & Gyurkovics, E. (2007). Sampled-data model predictive control for nonlinear time-varying systems: Stability and robustness, *in* R. Findeisen, F. Allgöwer & L. Biegler (eds), *Assessment and Future Directions of Nonlinear Model Predictive Control*, LNCIS 358, Springer, Berlin, pp. 115–129.

Graichen, K., Egretzberger, M. & Kugi, A. (2010). Suboptimal model predictive control of a laboratory crane, *8th IFAC Symposium on Nonlinear Control Systems (NOLCOS 2010)*, Budapest (Hungary).

Graichen, K., Kiefer, T. & Kugi, A. (2009). Real-time trajectory optimization under input constraints for a flatness-controlled laboratory helicopter, *European Control Conference (ECC) 2009*, Budapest (Hungary), pp. 2061–2066.

Graichen, K. & Kugi, A. (2010). Stability and incremental improvement of suboptimal MPC without terminal constraints, *IEEE Transactions on Automatic Control* 55(11): 2576–2580.

Graichen, K., Treuer, M. & Zeitz, M. (2007). Swing-up of the double pendulum on a cart by feedforward and feedback control with experimental validation, *Automatica* 43(1): 63–71.

Grüne, L. & Pannek, J. (2011). *Nonlinear Model Predictive Control: Theory and Algorithms*, Springer, London.

Hsu, J. & Meyer, A. (1968). *Modern Control Principles and Applications*, McGraw-Hill, New York.

Ito, K. & Kunisch, K. (2002). Receding horizon optimal control for infinite dimensional systems, *ESAIM: Control, Optimisation and Calculus of Variations* 8: 741–760.

Jadbabaie, A., Yu, J. & Hauser, J. (2001). Unconstrained receding horizon control of nonlinear systems, *IEEE Transactions on Automatic Control* 46(5): 776–783.

Kothare, S. D. O. & Morari, M. (2000). Contractive model predictive control for constrained nonlinear systems, *IEEE Transactions on Automatic Control* 45(6): 1053–1071.

Lee, E. & Markus, L. (1967). *Foundations of Optimal Control*, Wiley, New York.

Lee, Y., Kouvaritakis, B. & Cannon, M. (2002). Constrained receding horizon predictive control for nonlinear systems, *Automatica* 38(12): 2093–2102.

Leese, S. (1977). Convergence of gradient methods for optimal control problems, *Journal of Optimization Theory and Applications* 21(3): 329–337.

Limon, D., Alamo, T., Salas, F. & Camacho, E. (2006). On the stability of constrained MPC without terminal constraint, *IEEE Transactions on Automatic Control* 51(5): 832–836.

Mayne, D., Rawlings, J., Rao, C. & Scokaert, P. (2000). Constrained model predictive control: stability and optimality, *Automatica* 36(6): 789–814.

Michalska, H. & Mayne, D. (1993). Robust receding horizon control of constrained nonlinear systems, *IEEE Transactions on Automatic Control* 38(11): 1623–1633.

Nikol'skii, M. (2007). Convergence of the gradient projection method in optimal control problems, *Computational Mathematics and Modeling* 18(2): 148–156.

Ohtsuka, T. (2004). A continuation/GMRES method for fast computation of nonlinear receding horizon control, *Automatica* 40(4): 563–574.

Parisini, T. & Zoppoli, R. (1995). A receding-horizon regulator for nonlinear systems and a neural approximation, *Automatica* 31(10): 1443–1451.

Primbs, J. (1999). *Nonlinear Optimal Control: A Receding Horizon Approach*, PhD thesis, California Institute of Technology, Pasadena, CA.

Rawlings, J. & Mayne, D. (2009). *Model Predictive Control: Theory and Design*, Nob Hill Publishing, Madison, WI.

Scokaert, P., Mayne, D. & Rawlings, J. (1999). Suboptimal model predictive control (feasibility implies stability), *IEEE Transactions on Automatic Control* 44(3): 648–654.

Steinböck, A., Graichen, K. & Kugi, A. (2011). Dynamic optimization of a slab reheating furnace with consistent approximation of control variables, *IEEE Transactions on Control Systems Technology* 16(6): 1444–1456.

2

Feedback Linearization and LQ Based Constrained Predictive Control

Joanna Zietkiewicz
Poznan University of Technology,
Institute of Control and Information Engineering,
Department of Control and Robotics,
Poland

1. Introduction

Feedback linearization is a powerful technique that allows to obtain linear model with exact dynamics (Isidori,1985), (Slotine & Li, 1991). Linear quadratic control is well known optimal control method and with its dynamic programming properties can be also easily calculated (Anderson & Moore, 1990). The combination of feedback linearization and LQ control has been used in many algorithms in Model Predictive Control applications for many years and it is used also in the current papers (He De-Feng et al.,2011), (Margellos & Lygeros, 2010). Another problem apart from finding the optimal solution on a given horizon (finite or infinite) is the constrained control. A method which uses the advantages of feedback linearization, LQ control and applying signals constraints was proposed in (Poulsen et al., 2001b). In every step it is based on interpolation between the LQ optimal control and a feasible solution - the solution that fulfils given constraints. A feasible solution is obtained by taking calculated from LQ method optimal gain for a perturbed reference signal. The compromise between the feasible and optimal solution is calculating by minimization of one variable - the number of degrees of freedom in prediction is reduced to one variable.

Feedback linearization relies on choosing new state input and variables and then compensating nonlinearities in state equations by nonlinear feedback. The signals from nonlinear system are constrained, they are accessible from linear model through nonlinear equations. Therefore in the interpolation a nonlinear numerical method has to be used. The whole algorithm is operating in a discretized system.

There are several problems while using the method. One of them is that signals from nonlinear system can change its values within given one discrete time interval, while we assume that variables of linear model are unchanged. Those values should be considered as constrained. Another problem is finding the basic feasible perturbed reference signal which will provide well control performance. Method proposed in (Poulsen et. al, 2001b) gives good results if the weight matrices in cost function and the sampling interval are well chosen. Often it is difficult to choose these parameters and in general the solution may provide not only unfeasible signals (violating constraints), but also signals which violate assumption for system equations (like assumption of nonzero values in a denominator of a fraction).

Other method of finding feasible solution proposed in the chapter provides better results of feasibility. The presented method also takes into consideration important feature, that input of nonlinear system changes its value in the sampling interval, while the control value of linearized model is unchanged. The algorithm is applied to the two tanks model and also to the continuous stirred tank reactor model, which operates in an area of unstable equilibrium point. The influence of well chosen perturbed reference signal is presented on charts for those two systems. The chapter is closed by concluding remarks.

2. Input–output feedback linearization

The main idea in feedback linearization is the assumption that the object described by nonlinear equations is not intrinsically nonlinear but may have wrongly chosen state variables or input. By nonlinear compensation in feedback and new variables one can obtain linear model with embedded original model and its dynamics. A nonlinear SISO model

$$\begin{aligned} \dot{x} &= f(x) + g(x)u \\ y &= h(x) \end{aligned} \tag{1}$$

has a linear equivalent

$$\begin{aligned} \dot{z} &= Az + Bv \\ y &= Cz \end{aligned} \tag{2}$$

if there exists a diffeomorphism

$$z = \varphi(x) \tag{3}$$

and a feedback law

$$u = \psi(v, x). \tag{4}$$

Important factor in feedback linearization is a relative degree. This value represents of how many times the output signal has to be differentiated as to obtain direct dependence on input signal. If relative degree r is definite for the system then there is a simple method of obtaining linear system (2) with order r. It can be developed by differentiating r times the output variable y and by choosing new state variables and input as

$$\begin{aligned} y &= z_1 \\ \dot{y} &= z_2 \\ y^{(r-1)} &= z_r \\ y^{(r)} &= v \end{aligned} \tag{5}$$

where the derivatives can also be expressed by Lee derivatives

$$\dot{y} = L_f h(x) = \frac{dh(x)}{dx} f(x),$$

$$y^{(r-1)} = L_f^{\,r-1} h(x) = \frac{dL_f^{\,r-1} h(x)}{dx} f(x),$$

$$y^{(r)} = L_f{}^r h(x) + L_g L_f{}^{r-1} h(x)u = \frac{dL_f{}^{r-1}h(x)}{dx} f(x) + \frac{dL_f{}^{r-1}h(x)}{dx} g(x)u.$$

The linear system (5) describes the dependence between the new input v and the output y. These equations can be used to design appropriate input v in order to receive desirable output y. If relative degree r is smaller than the order of original nonlinear system n, then to track all state variables x we need additional n-r variables z. For

$$\eta(x) = \begin{bmatrix} z_{r+1} & \cdots & z_n \end{bmatrix}^T \tag{6}$$

the variables from vector (6) should satisfy condition

$$L_g \eta(x) = 0. \tag{7}$$

In that case the system has internal dynamics which has to be taken into consideration in stability analysis. The convenient way to consider the stability of n-r variables which after linearization are unobservable from output y is the analysis the zero dynamic. The zero dynamics is the internal dynamics of the system when the output is kept at zero by input. By using appropriate input and state and then checking the stability of obtained equations it is possible to find out if the system is minimum phase and the unobservable from y variables will converge to a certain value when time tends to infinity.

Feedback linearization method (Isidori,1985), (Slotine & Li, 1991) in the basic version is restricted to the class of nonlinear models which are affine in the input and have smooth functions $f(x)$, $g(x)$, definite relative degree and stable zero dynamics. Therefore algorithms which uses feedback linearization are limited by above conditions.

3. Unconstrained control

Unconstrained LQ control will be applied to discrete system

$$\begin{aligned} z_{k+1} &= A_d z_k + B_d v_k \\ y_k &= C_d z_k \end{aligned} \tag{8}$$

obtained by feedback linearization of (1) and by discretization of (2) with sampling interval Ts.

In order to track the nonzero reference signal w_t we augment the state space system by adding new variable z_{int} with integral action

$$z_{\text{int}\_t+1} = z_{\text{int}\_t} + w_t - y_t \tag{9}$$

the equation (8) with augmented state vector takes form

$$\begin{aligned} z_{t+1} &= \begin{bmatrix} A_d & 0 \\ -C_d & 1 \end{bmatrix} z_t + \begin{bmatrix} B_d \\ 0 \end{bmatrix} v_t + \begin{bmatrix} 0 \\ 1 \end{bmatrix} w_t \\ y_t &= \begin{bmatrix} C_d & 0 \end{bmatrix} z_t \end{aligned} \tag{10}$$

The cost function can be written by

$$J_t = \sum_{k=t}^{\infty} z_k{}^T Q z_k + R v_k{}^2, \tag{11}$$

then the control law which minimize the cost function (11)

$$v_t = L_y w_t - L z_t, \tag{12}$$

where L is the optimal gain and $L_y = L\begin{bmatrix} C_d & 0 \end{bmatrix}^T$.

If the system (11) is complete controllable and the weight matrices Q and R are positive definite, then the cost function J_t is finite and the control law (12) guarantee stability of the control system (Anderson & Moore 1990).

4. Constrained predictive control

Constrained variables of nonlinear system (1) can be expressed by equation

$$c_k = P x_k + H u_k \tag{13}$$

with constraints vectors LB and UB

$$LB \le c_k \le UB. \tag{14}$$

Constraints will be included into control law by interpolation method in every step t. It operates by using optimal control law (12) to

- original reference signal w_t (unconstrained optimal control),
- changed reference signal $\tilde{w}_t = w_t + p_t$ with p_t called perturbation so chosen, that all signals after using control law will satisfy constraints,

then using $\tilde{w}_t = w_t + \alpha_t p_t$ one has to minimize in every step a_t with constraints (14) while using (10) and (12) to predict future values on prediction horizon. For nonlinear system constrained values depend on signals from linear model through nonlinear functions (3,4) therefore to minimize a_t the bisection method was used in simulations.

The a_t can take values between 0 (this represents unconstrained control) and 1 (feasible but not optimal solution). If changing control v_t have the effect in changing u and every constrained values in monotonic way then the dependence of a_t on constrained values is also monotonic and there exists one minimum of a_t.

Note that p_t is a vector of the size of reference signal w_t calculated in the time instant t. The perturbation p_t which provide feasible solution can be obtained from previous step by

$$p_t = \alpha_{t-1} p_{t-1.} \tag{15}$$

With optimal a_t we can rewrite control law from (12):

$$v_t = L_y(w_t + \alpha_t p_t) - L z_t \tag{16}$$

and the state equation (10) with used (16):

$$z_{t+1} = \Phi z_t + \Gamma(w_t + \alpha_t p_t), \tag{17}$$

where

$$\Phi = \begin{bmatrix} A_d - B_d\begin{bmatrix} L_1 & L_2 \end{bmatrix} & -B_d L_3 \\ -C_d & 1 \end{bmatrix}, \tag{18}$$

$$\Gamma = \begin{bmatrix} B_d L_y \\ 1 \end{bmatrix}. \tag{19}$$

At the beginning of the algorithm (t=0) we have to find p_t in other way - we do not have p_{t-1}. Several ways of choosing this initial perturbation p_0 will be presented with analysis of its performance in the section 7.1.

5. Two coupled tanks

Equations describing dynamics of two tanks system

$$\begin{aligned} c\dot{h}_1 &= q - q_1 \\ c\dot{h}_2 &= q_1 - q_2 \end{aligned} \tag{20}$$

with Bernoulli equations

$$\begin{aligned} q_1 &= \sigma_l a_l \sqrt{2g(h_1 - h_2)} \quad for \quad h_1 \geq h_2 \\ q_2 &= \sigma_0 a_0 \sqrt{2gh_2} \quad for \quad h_2 \geq 0 \end{aligned} \tag{21}$$

presents action of the system. The variables h_1 and h_2 represent levels of a fluid in the first and the second tank. h_2 is also the output of the system. The control input is the inflow q to the first tank and the output is the level in the second tank. More details about this system can be find in (Poulsen et al.2001b).

After replacing the state by vector x and the input by u after some calculation we obtain system (1) with

$$\begin{aligned} f(x) &= \begin{bmatrix} \dfrac{-\sigma_l a_l}{c}\sqrt{2g(x_1 - x_2)} \\ \dfrac{\sigma_l a_l}{c}\sqrt{2g(x_1 - x_2)} - \dfrac{\sigma_0 a_0}{c}\sqrt{2gx_2} \end{bmatrix} \\ g(x) &= \begin{bmatrix} 1/c \\ 0 \end{bmatrix} \\ h(x) &= x_2. \end{aligned} \tag{22}$$

System inflow and the two levels are constrained in this system owing to its structure. Constrains are given by equations:

$$\begin{aligned} 0\text{cm}^3/\text{s} \leq u &\leq 96.3\text{cm}^3/\text{s} \\ 0\text{cm} \leq x_1 &\leq 60\text{cm} \\ 0\text{cm} \leq x_2 &\leq 60\text{cm}. \end{aligned} \tag{23}$$

5.1 Feedback linearization

By differentiating the output signal and choosing the consequent elements of vector z:

$$y = z_1 = x_2$$
$$\dot{y} = z_2 = L_f h(x)$$
$$\ddot{y} = \beta v = L_f{}^2 h(x) + L_g L_f h(x) u$$

we obtain linear system

$$\dot{z} = \begin{bmatrix} 0 & 1 \\ 0 & 0 \end{bmatrix} z + \begin{bmatrix} 0 \\ \beta \end{bmatrix} v \tag{24}$$

Where $\beta = 5 \times 10^{-5}$ is chosen to ensure balanced relation of components in LQ cost equation.

While operating on linear model we need to have access to state variables the diffeomorphism (3). We also need equation to calculate the control signal from original system (4).

This can be done via the following equations (calculated as a result of (24) and above):

$$x = \varphi(z) = \begin{bmatrix} z_1 + \dfrac{\left(cz_2 + \sigma_0 a_0 \sqrt{2gz_1}\right)^2}{2g\sigma_l^2 a_l^2} \\ z_1 \end{bmatrix} \tag{25}$$

$$u = \psi(v, x) = \frac{\beta v - L_f{}^2 h(x)}{L_g L_f h(x)} \tag{26}$$

6. Continuous stirred tank reactor

The operation of reactor (CSTR) is described by 3 differential equations (27). First equation illustrates the mass balance,

$$V\frac{dC(t)}{dt} = \phi(C_i - C(t)) - VR(t), \tag{27a}$$

where $C(t)$ is the concentration (molar mass) of reaction product measured in [kmol/m^3]. The second equation represents the balance of energy in the reactor

$$V\rho c_p \frac{dT(t)}{dt} = \phi\rho c_p (T_i - T(t)) - Q(t) + \delta VR(t), \tag{27b}$$

the balance of energy in the reactor cooling jacked is described by third equation

$$v_j \rho_j c_{pj} \frac{dT_j(t)}{dt} = \phi_j(t) \rho_j c_{pj} \left(T_{j0} - T_j(t)\right) + Q(t), \tag{27c}$$

with $T(t)$ - temperature inside the reactor and $T_j(t)$ – temperature in the cooling jacket, both measured in Kelvin.

Thermal energy in the process of cooling and the velocity of reaction are described by additional equations:

$$Q(t) = UA_c\left(T(t) - T_j(t)\right),$$

$$R(t) = C(t)k_0e^{-E/RT(t)}.$$

$\phi_j(t)$ represents cooling flow through the reactor jacket expressed in [m^3 /h] and is the input of the system. The output variable is the temperature $T(t)$. More detailed explanation of this system can be found in (Zietkiewicz, 2010).

Equations (27) can be rearranged to the simplified form (1) with

$$f(x) = \begin{bmatrix} aC_i - (a + k_0e^{-E/Rx_2})x_1 \\ aT_i - (a + b_1)x_2 + b_1x_3 + cx_1k_0e^{-E/Rx_2} \\ b_2(x_2 - x_3) \end{bmatrix}$$

$$g(x) = \begin{bmatrix} 0 \\ 0 \\ \dfrac{T_{j0} - x_3}{v_j} \end{bmatrix} \quad (28)$$

$$h(x) = x_2,$$

where

$$a = \frac{\phi}{V},\ b_1 = \frac{UA_c}{V\rho c_p},\ b_2 = \frac{UA_c}{v_j\rho_j c_{pj}},\ c = \frac{\delta}{\rho c_p}.$$

Constrained value in this system is the inflow of the cooling water to the reactor jacket -the input of the system

$$0\text{m}^3/\text{h} \le u \le 2.5\text{m}^3/\text{h} \quad (29)$$

The system has an interesting property – three equilibrium points, two stable and one unstable. In normal work the system is operating in the unstable area.

6.1 Feedback linearization

The system has order n=3 relative degree r=2. Therefore we obtain two linear equations (two states) differentiating the output

$$y = z_1 = x_2$$
$$\dot{y} = z_2 = L_f h(x)$$
$$\ddot{y} = \beta v = L_f{}^2 h(x) + L_g L_f h(x)u$$

We obtain linear system with order=2 similar to (24). The calibrating parameter in this system $\beta = 5\times10^{-4}$. The system has internal dynamic described by equation

$$\dot{x}_1 = aC_i - (a + k_0 e^{-E/Ry})x_1$$

The zero dynamics are given by

$$\dot{x}_1 = aC_i - ax_1$$

The eigenvalue is then equal to a. As $\phi = 1.13\text{m}^3 / h$ and $V = 1.36\text{m}^3$ the modulus of a is less than 1 therefore the system is minimum phase.

The third state variable satisfying condition (7) will be chosen as

$$z_3 = x_1,$$

then

$$x = \varphi(z) = \begin{bmatrix} z_3 \\ z_1 \\ \dfrac{(a+b_1)z_1 + z_2 - cz_3 k_0 e^{-E/Rz_1} - aT_i}{b_1} \end{bmatrix}, \tag{30}$$

$$u = \psi(v,x) = \frac{\beta v - L_f{}^2 h(x)}{L_g L_f h(x)}. \tag{31}$$

7. Operating of the algorithm

The control strategy described in sections 2-4 will be developed in this point showing advantages of the algorithm while using it to the two nonlinear systems with constraints.

7.1 Initial perturbation

Problem with finding initial perturbation signalized at the end of the section 4, arise because the solution must guarantee constraints, and the constrained values in spite of linearization are not accessible in a linear way. On the other hand this solution should not be too simple and only feasible as it will be shown on charts.

The first way of calculating initial perturbation is the method proposed in (Poulsen et al.2001b). It is based on using zero as the reference signal and the initial state corresponding to the step of original reference signal. We obtain state equation

$$z_{t+1} = \Phi z_t + \Gamma p_t. \tag{32}$$

After minimization of the cost function

$$J_t = \sum_{k=t}^{\infty} z_k{}^T Q_p z_k + R_p p_k{}^2 \tag{33}$$

and finding optimal gain K by LQ method we have

$$p_t = -Kz_t. \tag{34}$$

In fig.(1) charts with dashed lines presents signals without perturbation and with zero reference signal, whereas solid lines represent signals with used perturbation obtained from (34). Minimization of the first element in (33) approaches output and input v to zero, minimization of the second element approaches signals to that without using perturbation. Problem appears with the input v which approaches to zero by minimization of the first element of (33) but by minimization of the second element approaches to high negative value. This is visible in the first steps. This value also depends on Q_p and R_p nonetheless it cannot be chosen arbitrarily close to zero. Too high modulus of v causes signals of nonlinear system to be more didstant from zero, and that can violate constraints. Another way of calculating initial perturbation can be find in (Poulsen et al.2001a) but that method is limited to linear (or Jacobian linearized) models.

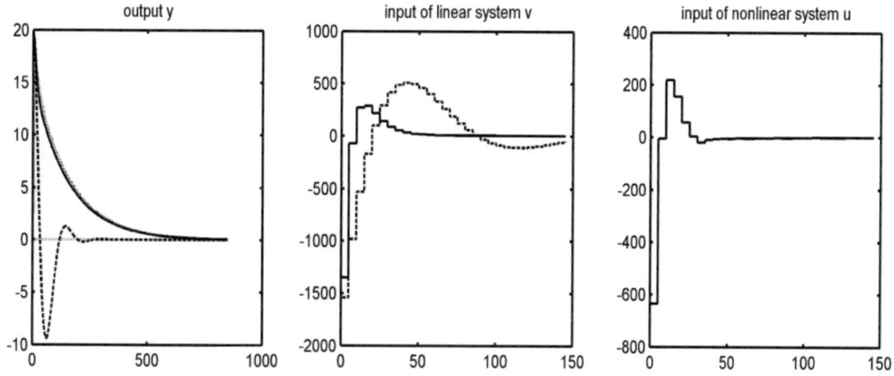

Fig. 1. First method of finding the initial perturbation trajectory

To remedy this difficulty we can try to use as the initial perturbation signal which makes w_t and automatically other signals unchanged. This however causes problems in working algorithm in next steps and provides week tracking of original reference signal (this will be shown in fig.(11)).

Other way of calculating initial perturbation is to take minimum of

$$J_t = \sum_{k=t}^{\infty} z_k^T Q_p z_k + R_p v_k^2 \tag{35}$$

when

$$v_t = -Lz_t + L_y p_t \tag{36}$$

then after some calculations

$$J_t = \sum_{k=t}^{\infty} z_k^T Q_j z_k + R_p p_k^2 + 2z_k^T N_j p_k \tag{37}$$

with

$$Q_j = Q_p + L^T RL, \quad R_j = L_y RL_y, \quad N_j = -L^T RL_1 \tag{38}$$

After using this cost function (37) with the same Q_p and R_p as was used in the first method of calculating initial perturbation we obtain signals presented in fig.(2).

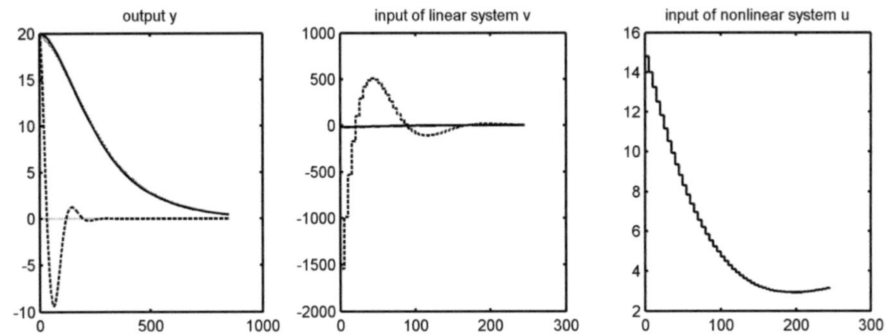

Fig. 2. Second method of finding the initial perturbation trajectory

It can be seen from figures (1) and (2) that in the second variant the two input values have smaller absolute values which can have an influence on fulfilling constraints. The second solution is not provide feasible signals for every Q, R, Q_p R_p, T_s but it simplify choosing those parameters.

7.2 Constrained values as a dependence of α

After using the third method of obtaining initial perturbation for model of two tanks and reactor we will see how the constrained values are dependent on a_t in the first step. Important feature of nonlinear system is that in a sampling interval T_s in given step t when v_t is constant, u is changing because u is a function of v_t and x, which is also changing from x_t to x_{t+1}. We have to monitor this control value as it may violate constraints. We can calculate x in every step from the inversion of (3) but (4) gives as only initial u_t at the beginning of T_s. Therefore u has to be calculated by integration. However when T_s is not to high and u changes monotonically in T_s we can use its approximated value at the end of T_s calculated from (4) by

$$u_{t\_end} = \psi(v_t, x_{t+1}). \tag{39}$$

That value has to be taken in consideration in the algorithm while minimizing a_t with constraints.

For the two tanks system we have constrained u, x_1 and x_2. Constraints are given in (23).

Figures represent how the input and the two variables change for various a_t. The system was sampled with T_s=5, weight matrices for LQ regulator are given Q=diag(1 1 1), R=0.01 and the weight matrices used to calculate initial perturbation are Q_p=0.01* diag(1 1 1), R_p=1. Reference signal was changed from 20cm to 40cm.

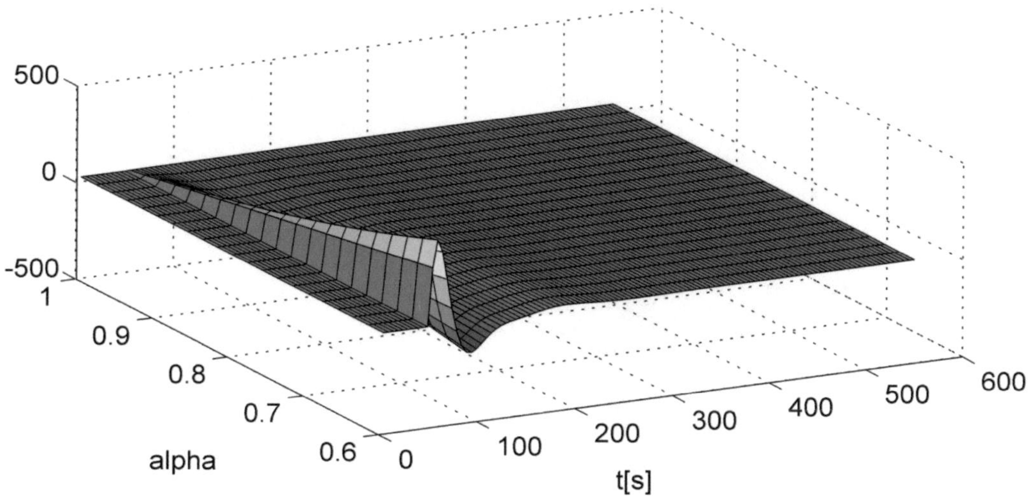

Fig. 3. Input u[cm³/s] as a dependence on a

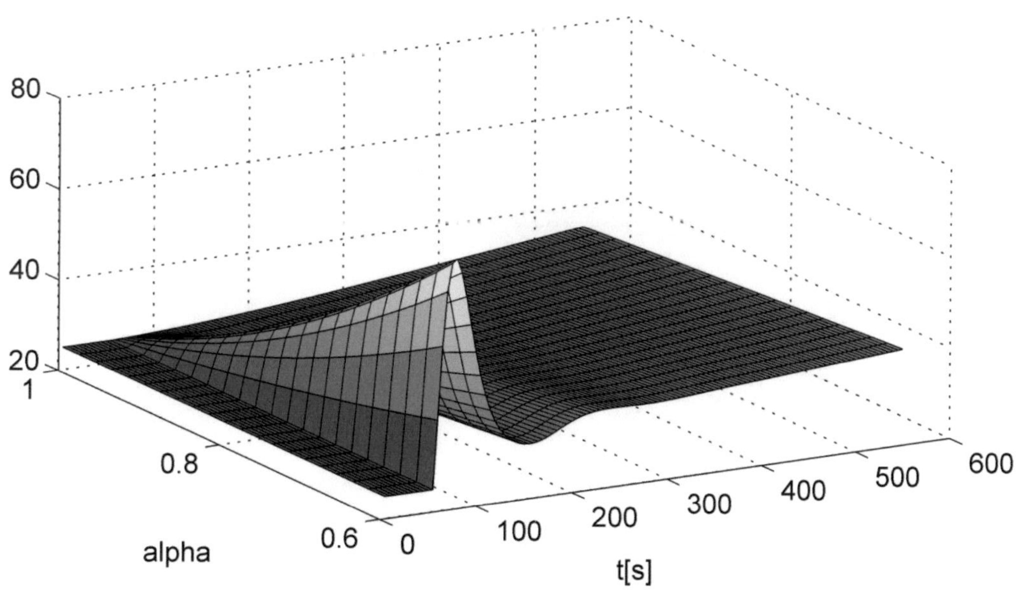

Fig. 4. Level in the first tank x_1[cm] as a dependence on a

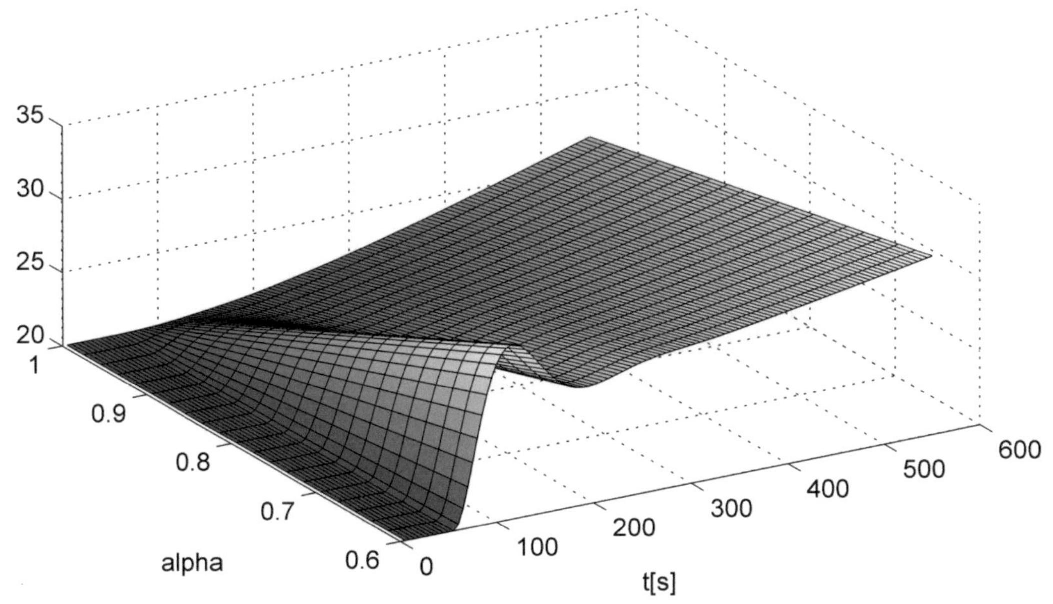

Fig. 5. Level in the second tank x_2[cm] as a dependence on a

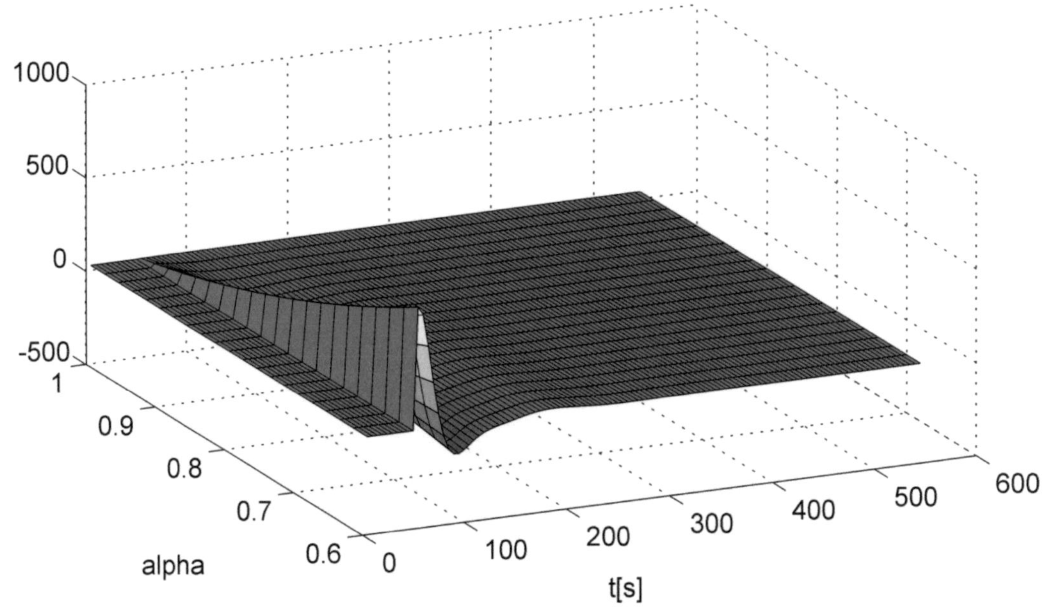

Fig. 6. Input u[cm^3/s] calculated at the end of every T_s as a dependence on a

On above figures it can be seen that the dependence of x and u on a_t is monotonic and for small values a_t the variables are close to zero end fulfils constraints. We can see that input values at the end of every period T_s is very important because it can takes higher values than u_t calculated from (4).

The CSTR system has one constrained value - control input u, the constraints are given in equation (29). For simulations the sampling interval was chosen as T_s=5s, weight matrices for LQ regulator: Q=diag(1 1 1), R=10 and weight matrices for LQ regulator in first perturbation calculations: Q_p=0.1*diag(1 1 1), Rp=10. Reference values was changed from 333K to 338K.

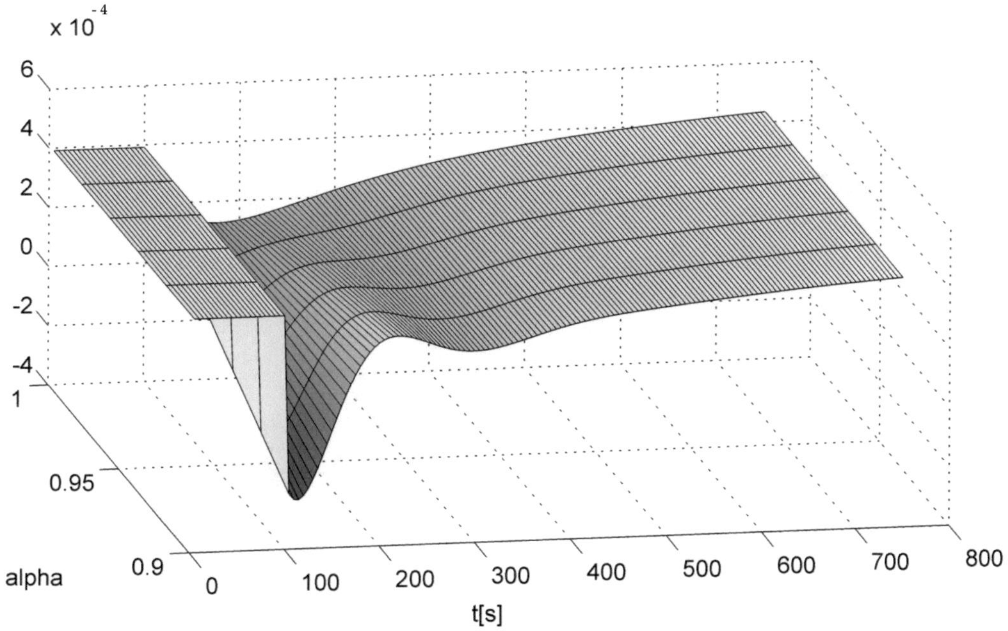

Fig. 7. Input u[m³/h] as a dependence on a

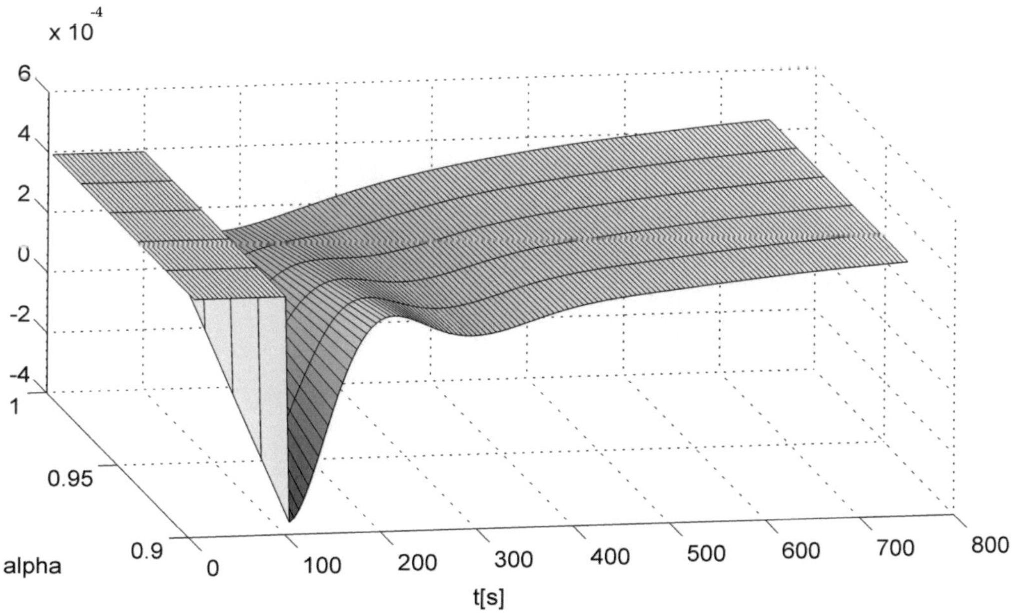

Fig. 8. Input u[m³/h] calculated at the end of every T_s as a dependence on a

In figures (7-8) we can see as for the two tank system that constrained values are monotonically dependent on a. Moreover the two unconstrained variables x_1 and x_2 which charts are presented in fig.(9,10) are also monotonically dependent on a therefore those variables could be taken into consideration as constrained variables in the algorithm.

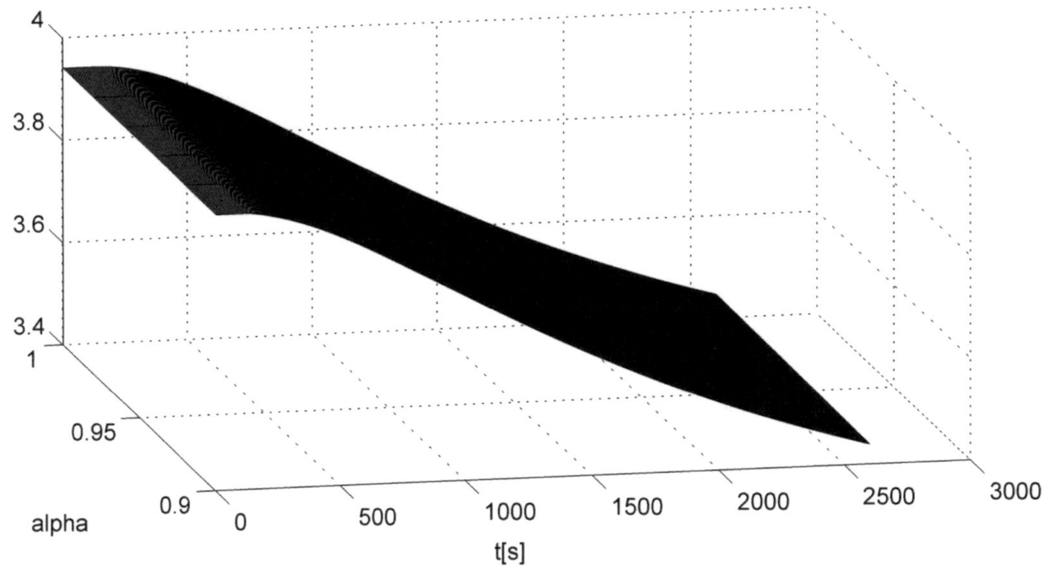

Fig. 9. Product concentration x_1[kmol/m^3] as a dependence on a

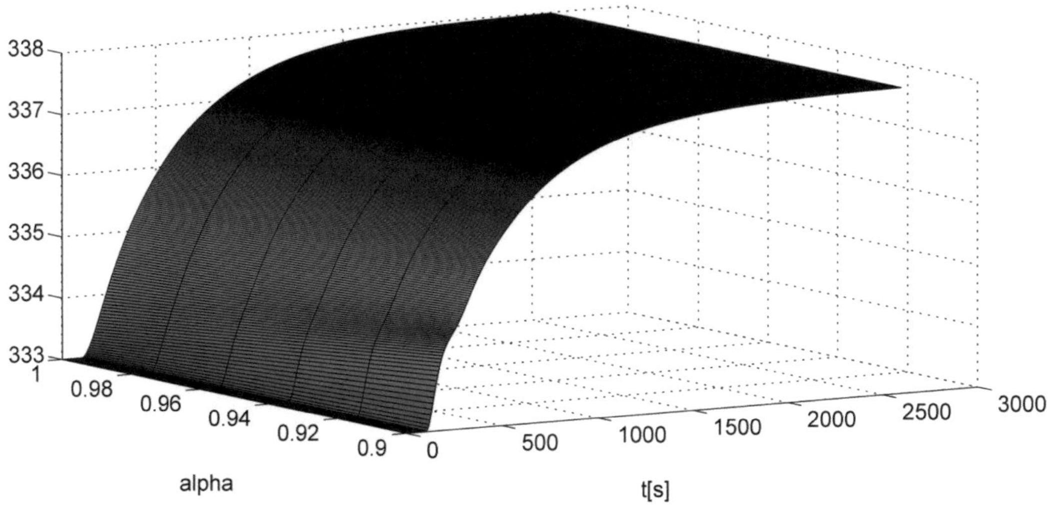

Fig. 10. Temperature in the jacket x_2[K] as a dependence on a

7.3 Simulations of the algorithm

In this section the final algorithm is used for two tanks system and then for CSTR system. On every figure time is expressed in seconds. For the two tanks system reference signal was changed from 20cm to 40cm in time 160s, other adjustments were chosen as: T_s=8s, Q=diag(1 1 1), R=0.1.

In the first experiment the initial perturbation was chosen so that reference signal and therefore every signals in the system was unchanged. The result is given in fig.(11).

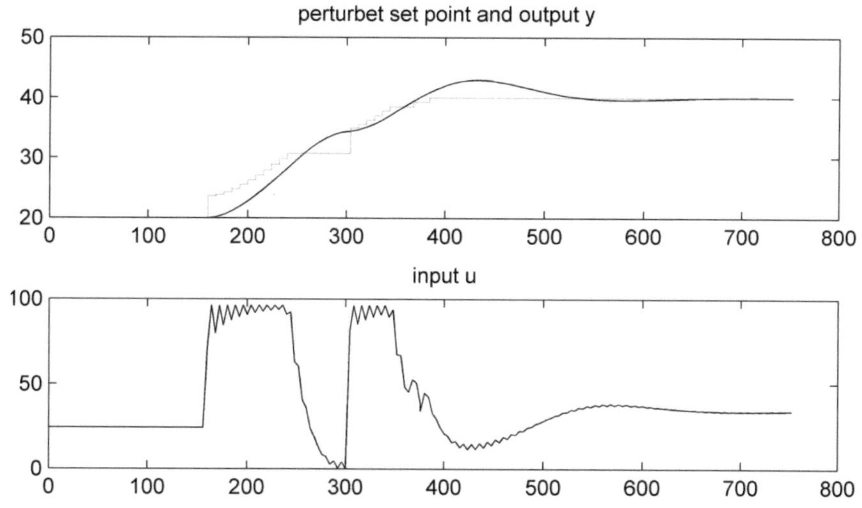

Fig. 11. First experiment for two tanks system, output y[cm] and input u[cm^3/s] values

In this case if we use perturbed reference trajectory obtained in the described way, in every time instant t changing a_t means that the perturbed reference signal is a step in this time instant and it is not changing from time t+1 to the end of original reference signal. In the upper chart the output is represented by solid line, whereas dotted line means perturbed reference signal (the first value of the perturbed reference signal is taken in every step t). There is visible that from about 250s to 300s the perturbation is the same, in those instants a has to be equal 1. That is a consequence of too low perturbed reference signal which results in too low value of input, which has to be placed by appropriate a at the constraint, in this case zero. In normal work of this algorithm if the active constraint is the constraint of input it should concern values in the first steps distant from the current t.

In the second experiment we will use initial perturbation calculated with cost function (37) and weight matrices Q_p=0.1*diag(1 1 1), R_p=0.1.

In the second experiment the active constraint is the input and from time 270s the level in the first tank. The regulation time is shorter than in the first experiment, constraints are fulfilled. The fast changes of input value visible from time 150s are the changes within intervals T_s.

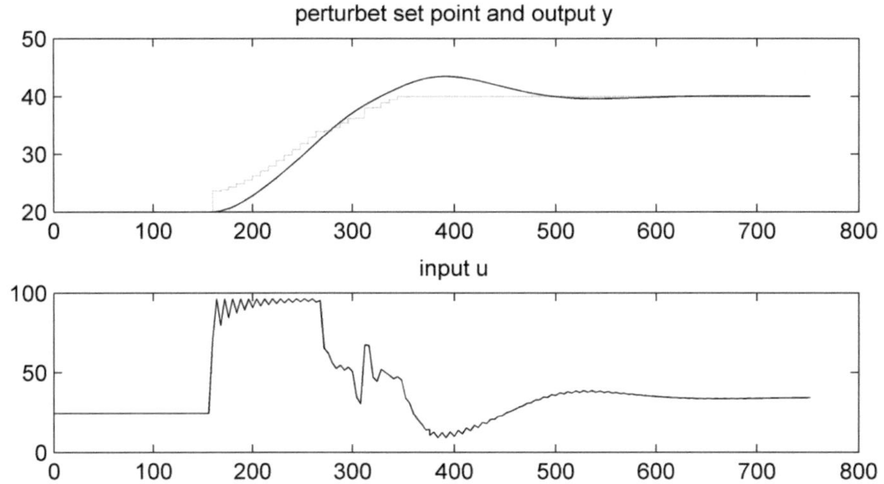

Fig. 12. Second experiment for two tanks system, output y[cm] and input u[cm^3/s] values

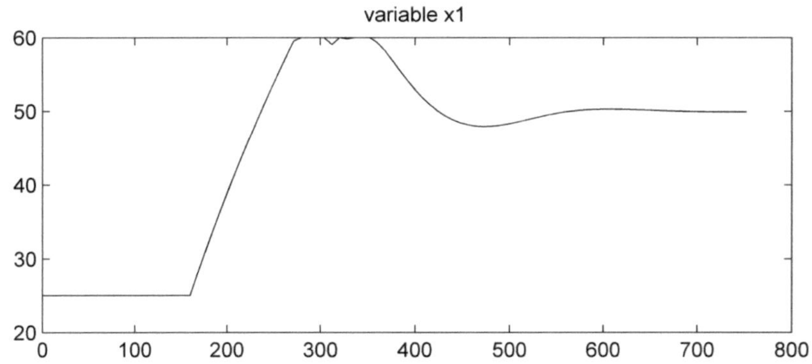

Fig. 13. The level in the first tank x_1[cm] in the second experiment for two tanks system

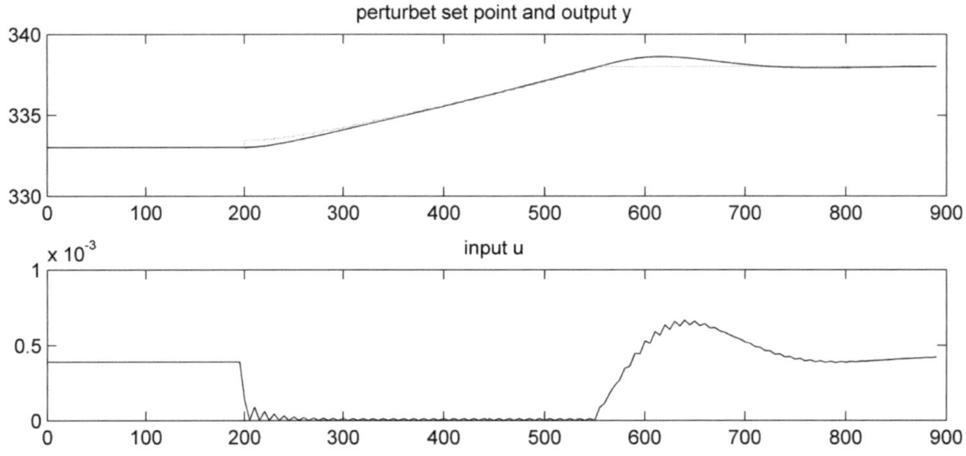

Fig. 14. The experiment for the CSTR system, output y[K] and input u[m^3/h] values

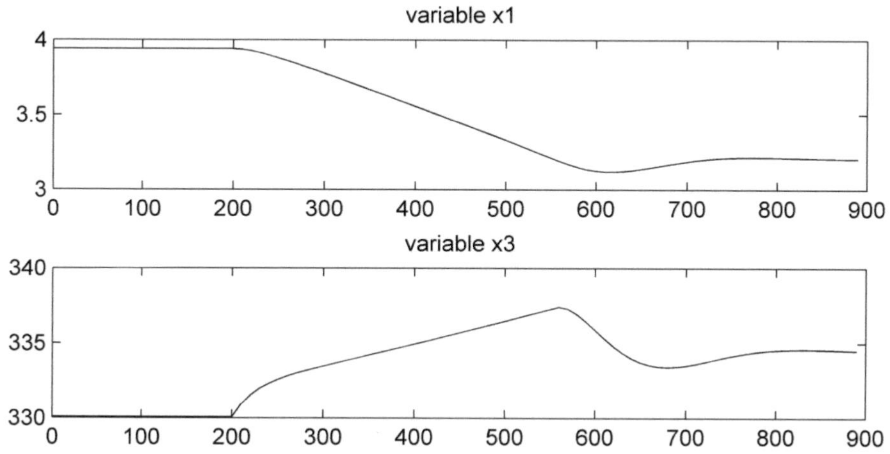

Fig. 15. The experiment for the CSTR system, product concentration x_1[kmol/m^3] and the temperature in the jacket x_2[K]

The experiment for Continuous Stirred Tank Reactor was performed for changing reference signal from 333K to 338K, adjustments takes given values: T_s=10, Q=diag(1 1 1), R=10 Q_p=0.1*diag(1 1 1), R_p=10.

8. Conclusion

Model based predictive control attracts interest of researchers for many years as the method which is intuitive and allows to include constraints in the control design. Quadratic cost function in various types are used in MPC. Application of feedback linearization in MPC is also interested issue. Proposed interpolation method allows to reducing the number of degrees of freedom in the prediction. horizon. In the chapter the algorithm which combine interpolation and LQ regulator for feedback linearized system was tested for a CSTR model which is nonlinear and works in unstable area. It has been developed by using new initial perturbation calculating and by taking into consideration input values of unconstrained model which changes within sampling intervals.

Further research in this area could concern developing a method of finding adjustments for initial perturbation and for the LQ regulator used in the algorithm. Interesting issue is to apply the method for more complicated system. The multi-input and multi-output systems can be interesting class because feedback linearization rearranges those systems to m linear single-input, single output systems.

9. References

Anderson, B. D.O.; Moore J. B. Optimal control. Linear quadratic methods (1990), Prentice-Hall, ISBN 0-13-638560-5, New Jersey, USA

He De-Feng, Song Xiu-Lan, Yang Ma-Ying, (2011), *Proceedings of 30th Chinese Control Conference*, ISBN: 978-1-4577-0677-6, pp. 3368 – 3371, Yantai, China

Isidori A. (1985). Lecture Notes in Control and Information Sciences, Springer-Verlag, ISBN 3-540-15595-3, ISBN 0-387-15595-3, Berlin, Germany

Margellos, K.; Lygeros, J. (2010), *Proceedings of 49th IEEE Conference on Decision and Control,* ISBN 978-1-4244-7745-6, Atlanta, GA

Poulsen, N. K.; Kouvaritakis, B.; Cannon, M. (2001a). Constrained predictive control and its application to a coupled-tanks apparatus, International Journal of Control, pp. 74:6, 552-564, ISSN 1366-5820

Poulsen, N. K.; Kouvaritakis, B.; Cannon, M. (2001b). Nonlinear constrained predictive control applied to a coupled-tanks apparatus, IEE Proc. Of Control Theory and Applications, pp.17-24, ISNN 1350-2379

Slotine, J. E. ;Li W. (1991). Applied Nonlinear Control, Prentice-Hall, ISBN 0-13-040049-1, New Jersey, USA

Zietkiewicz, J. (2010), Nonlinear constrained predictive control of exothermic reactor, *Proceedings of 7th International Conference on Informatics in Control, Automation and Robotics*, ISBN 978-989-8425-02-7, Vol.3, pp.208-212, Funchal, Portugal

3

Infeasibility Handling in Constrained MPC

Rubens Junqueira Magalhães Afonso and
Roberto Kawakami Harrop Galvão
Instituto Tecnológico de Aeronáutica
Brazil

1. Introduction

1.1 Aim of the chapter

Predictive Control optimization problems may be rendered infeasible in the presence of constraints due to model-plant mismatches, external perturbations, noise or faults. This may cause the optimizer to issue a control sequence which is impossible to implement, leading to prediction errors, as well as loss of stability of the control loop. Such a problem motivates the development of techniques aimed at recovering feasibility without violating hard physical constraints imposed by the nature of the plant. Currently, setpoint management approaches and techniques dealing with changes in the constraints are two of the most effective solutions to recover feasibility with low computational demand. In this chapter a review of techniques that can be understood as one of the aforementioned is presented along with some illustrative simulation examples.

1.2 Concepts and literature review

One of the main advantages of Predictive Control is the ability to deal with constraints over the inputs and states of the plant in an explicit manner, which brings better performance and more safety to the operation of the plant (Maciejowski, 2002), (Rossiter, 2003). Constraints over the excursion of the control signals are particularly common in processes that operate near optimal conditions (Rodrigues & Odloak, 2005). However, if the optimization becomes infeasible, possibly due to model-plant mismatches, external perturbations, noise or faults, a control sequence which is impossible to implement may be issued, leading to prediction errors, as well as loss of stability of the control loop (Maciejowski, 2002). Such a problem motivates the development of techniques aimed at recovering feasibility without violating hard physical constraints imposed by the nature of the plant.

The MPC formulation itself allows for a simple solution, which consists of enlarging the horizons, as means to allow for more degrees of freedom in the optimization. On the other hand, an increase in the computational burden associated to the solution of the optimization problem results, since there are more decision variables as well as constraints. Moreover, enlarging the horizons cannot solve all sorts of infeasibilities.

Constraint relaxation is one alternative which involves less decision variables and is usually effective. Nevertheless, it is often not obvious which constraints to relax and the amount by which they should be relaxed in order to attain a feasible optimization problem. There are different approaches for this purpose, some of which will be briefly discussed in this chapter. Initially, one must differentiate between two types of constraints (Alvarez & de Prada, 1997), (Vada et al., 2001):

Physical constraints: those limits that can never be surpassed and are determined by the physical functioning of the system. For instance, a valve cannot be opened more than 100% or less than 0%.

Operating constraints: those limits fixed by the plant operator. These limits, which are usually more restrictive than the physical constraints, define the band within which the variables are expected to be under normal operating conditions. For instance, it may be more profitable to operate a chemical reactor in a certain range of temperatures, in order to favor the kinetics of the desired reaction that forms products of economical interest. However, if maintaining such operating condition would compromise the safety of operation of the plant at some point, then the associated constraints could be relaxed.

The literature has many different approaches to constraint relaxation. Some infeasibility handling techniques are described in Rawlings & Muske (1993) and Scokaert & Rawlings (1999):

Minimal time approach: An algorithm identifies the smallest time, $\kappa(x)$, which depends on the current state x, beyond which the state constraint can be satisfied over an infinite horizon. Prior to time $\kappa(x)$, the state constraint is ignored, and the control law enforces the state constraint only after that time. An advantage of this method is that it leads to the earliest possible constraint satisfaction. Transient constraint violations, however, can be large.

Soft-constraint approach: Violations of the state constraints are allowed, but an additional term is introduced in the cost function to penalize the constraint violation.

In Zafiriou & Chiou (1993) the authors propose a method for calculating the smallest magnitude of the relaxation that renders the optimization feasible for a SISO system.

The paper by Scokaert (1994) presents many suggestions to circumvent the problem of infeasibility, among which, one that classifies the constraints in priority levels and tries to enforce the ones with higher priority through relaxation of the others.

Scokaert & Rawlings (1999) introduce an approach capable of minimizing the peak and duration of the constraint violation, with advantages concerning the transient response.

A relaxation procedure that can be applied either to the controls or to the system outputs is described by Alvarez & de Prada (1997). The control-related approach consists of relaxing the operating constraints on the control amplitude or rate of change according to a priority schedule. The output-related approach consists of relaxing the operating constraints on the output amplitude or modifying the time interval where such constraints are imposed within the prediction horizon.

In Vada et al. (2001) the proposed scheme involves the classification of the constraints in priority levels and the solution of a linear programming problem parallel to the MPC optimization. In Afonso & Galvão (2010a), different weights are employed for the relaxation of operating output constraints, up to the values of physical constraints, as means to overcome infeasibility caused by actuator faults.

Another alternative to recover feasibility are the so-called setpoint management procedures (Bemporad & Mosca, 1994), (Gilbert & Kolmanovsky, 1995), (Bemporad et al., 1997), which artificially reduce the distance between the actual plant state and the constraint set. The reference governor proposed by Kapasouris et al. (1988) inspired many techniques to deal with problems involving actuator saturation through manipulation of the setpoint or the tracking error (Gilbert & Kolmanovsky, 1995). There are also papers aiming at imposing a reference model to the behavior of the plant that employ setpoint management in order to obtain feasibility when the control signals are bounded (Montandon et al., 2008).

Stability guarantees may be achieved with setpoint management by using a terminal constraint invariant set parameterized by the setpoint. Limon et al. (2008) employ this technique parameterizing the terminal set in terms of the control and state setpoints. The authors show that an optimal management of the setpoint may be achieved, guaranteeing the smallest distance between the desired setpoint and the one used by the MPC. This procedure increases the domain of attraction of the controller dramatically.

An application of the parameterization of the terminal set in terms of the steady-state value of the control can be found in Almeida & Leissling (2010). In that paper, the technique is employed to circumvent infeasibility caused by actuator faults which limit the range of values of control that the actuator can deploy. On the other hand, in Afonso & Galvão (2010b) the authors manage the setpoint of a state variable that does not affect the control setpoint, making parameterization of the terminal set unnecessary, as means to overcome infeasibility brought about by similar actuator faults.

In this chapter, the treatment of infeasibility in the optimization problem of constrained MPC will be discussed. Some illustrative simulations will provide a basic coverage of this topic, which is of great importance to practical implementations of MPC due to the capability of circumventing problems brought about by model-plant mismatch, faults, noise, disturbances or simply reducing the computational burden required to calculate an adequate control sequence.

2. Adopted MPC formulation

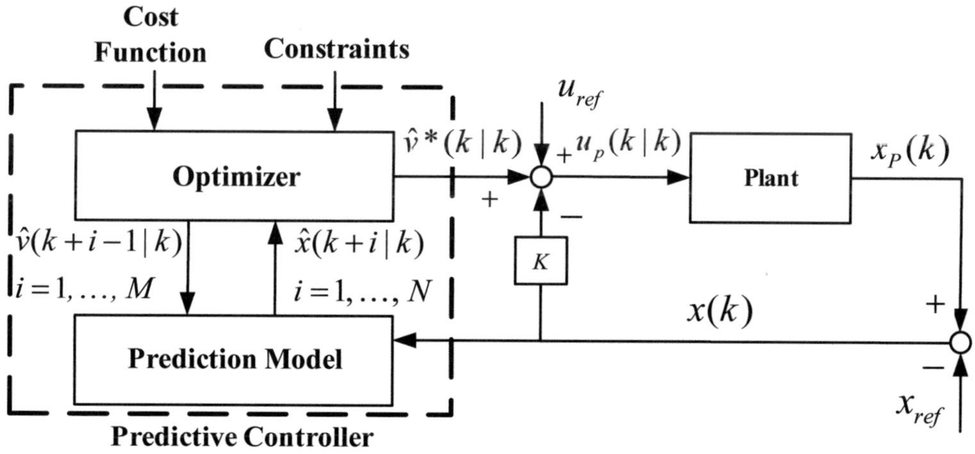

Fig. 1. MPC with inner feedback loop.

Fig. 1 presents the main elements of the MPC formulation adopted in this chapter. Since this is a regulator scheme, the desired equilibrium value x_{ref} for the state must be subtracted from the measured state of the plant x_P, in order to generate the state x read by the controller:

$$x = x_P - x_{ref} \tag{1}$$

In a similar manner, the corresponding equilibrium value of the control u_{ref} must be added to the output of the controller u to generate the control u_P to be applied to the plant, that is:

$$u = u_P - u_{ref} \tag{2}$$

A mathematical model of the plant is employed to calculate state predictions N steps ahead, over the so-called "Prediction Horizon". These predictions are determined on the basis of the current state ($x(k) \in \mathbb{R}^n$) and are also dependent on the future control sequence. $\hat{\bullet}(k+i|k)$ denotes the predicted value of variable $\bullet$ at time $k+i$ ($i \geq 1$) based on the information available at time k. The optimization algorithm determines a control sequence, over a Control Horizon of M steps ($\hat{v}(k+i-1|k)$, $i = 1, \ldots, M$), that minimizes the cost function specified for the problem, possibly subject to state and/or input constraints. It is assumed that the MPC control sequence is set to zero after the end of the Control Horizon, i.e. $\hat{v}(k+i-1|k) = 0$, $i > M$. The control is implemented in a receding horizon fashion, i.e., only the first element of the optimized control sequence is applied to the plant and the solution is recalculated at the next sampling period taking into account the new sensor readings. Therefore, the controller output at time k is given by $u(k) = \hat{u}^*(k|k) = \hat{v}^*(k|k) - Kx(k)$, where K is the gain of an internal loop.

It is assumed that the dynamics of the plant can be described by a discrete state-space equation of the form $x_P(k+1) = Ax_P(k) + Bu_P(k)$. Therefore, the relation between u and x is given by

$$x(k+1) = Ax(k) + Bu(k) \tag{3}$$

The MPC controller is designed to enforce constraints of the type

$$u_{P,min} \leq u_P \leq u_{P,max} \tag{4}$$

$$x_{P,min} \leq x_P \leq x_{P,max} \tag{5}$$

Considering Eqs. (1) and (2), the constrains in Eqs. (4) and (5) can be expressed as

$$u_{P,min} - u_{ref} \leq u \leq u_{P,max} - u_{ref} \tag{6}$$

$$x_{P,min} - x_{ref} \leq x \leq x_{P,max} - x_{ref} \tag{7}$$

The optimization problem to be solved at instant k consists of minimizing a cost function of the form

$$J_{mpc} = \sum_{i=0}^{M-1} \hat{v}^T(k+i|k)\Psi\hat{v}(k+i|k) \tag{8}$$

subject to the following constraints:

$$\hat{u}(k+i|k) = -K\hat{x}(k+i|k) + \hat{v}(k+i|k),\ i \geq 0 \tag{9}$$

$$\hat{v}(k+i|k) = 0,\ i \geq M \tag{10}$$

$$\hat{x}(k+i+1|k) = A\hat{x}(k+i|k) + B\hat{u}(k+i|k),\ i \geq 0 \tag{11}$$

$$\hat{x}(k|k) = x(k) \tag{12}$$

$$\hat{y}(k+i|k) = C\hat{x}(k+i|k),\ i \geq 0 \tag{13}$$

$$\hat{u}(k+i|k) \in \mathbb{U},\ i \geq 0 \tag{14}$$

$$\hat{x}(k+i|k) \in \mathbb{X},\ i > 0 \tag{15}$$

in which $\Psi = \Psi^T > 0$ is a weight matrix and $\mathbb{U}$ and $\mathbb{X}$ are the sets of admissible controls and states, respectively, according to Eqs. (6) and (7).

Following a receding horizon policy, the control at the k-th instant is given by $u(k) = \hat{v}^*(k|k) - Kx(k)$, where K is the gain of the internal loop represented in Fig. 1. At time $k+1$, the optimization is repeated to obtain $v^*(k+1|k+1)$.

The inner-loop controller is designed as a Linear Quadratic Regulator (LQR) with the following cost function:

$$\begin{gathered} J_{lqr} = \Sigma_{i=0}^{\infty} \left[\hat{x}^T(k+i|k)Q_{lqr}\hat{x}(k+i|k) + \hat{u}^T(k+i|k)R_{lqr}\hat{u}(k+i|k)\right], \\ Q_{lqr} = Q_{lqr}^T \geq 0, \quad R_{lqr} = R_{lqr}^T > 0 \end{gathered} \tag{16}$$

with Q_{lqr} chosen so that the pair $(A, Q_{lqr}^{\frac{1}{2}})$ is detectable.

Let P be the only non-negative symmetric solution of the Algebraic Riccati Equation $P = A^TPA - A^TPB(R_{lqr} + B^TPB)^{-1}B^TPA + Q_{lqr}$. It can then be shown that, if the weight matrix Ψ is chosen as $\Psi = R_{lqr} + B^TPB$, then the minimization of the cost in Eq. (8) subject to the constraints of Eqs. (9) – (15) is equivalent to the minimization of the cost of Eq. (16) subject to the constraints of Eqs. (11) – (15) (Chisci et al., 2001). The outcome is that the cost function has an infinite horizon, which is useful for stability guarantees (Scokaert & Rawlings, 1998), (Kouvaritakis et al., 1998). It is worth noting that, due to the penalization of the control signal $\hat{v}$ in the cost of Eq. (8), the MPC acts only when it is necessary to correct the inner-loop control in order to avoid violations of the constraints stated in Eqs. (14) and (15).

Defining vector $\hat{V}$ and matrix $\overline{\Psi}$ as

$$\hat{V} = \begin{bmatrix} \hat{v}(k|k) \\ \vdots \\ \hat{v}(k+M-1|k) \end{bmatrix}, \quad \overline{\Psi} = \begin{bmatrix} \Psi & \dots & 0 \\ \vdots & \ddots & \vdots \\ 0 & \dots & \Psi \end{bmatrix}, \tag{17}$$

the cost function can be rewritten as

$$J_{mpc} = \hat{V}^T\overline{\Psi}\hat{V} \tag{18}$$

which is quadratic in terms of $\hat{V}$.

Defining the vectors

$$\hat{X} = \begin{bmatrix} \hat{x}(k+1|k) \\ \vdots \\ \hat{x}(k+N|k) \end{bmatrix}, \quad \hat{U} = \begin{bmatrix} \hat{u}(k|k) \\ \vdots \\ \hat{u}(k+N-1|k) \end{bmatrix}, \tag{19}$$

the state and control prediction vectors may be related to $\hat{V}$ as (Maciejowski, 2002):

$$\begin{aligned} \hat{X} &= H\hat{V} + \Phi x(k) \\ \hat{U} &= H_u\hat{V} + \Phi_u x(k) \end{aligned} \tag{20}$$

It is important to remark that the presence of an infinite number of constraints in Eqs. (14) and (15) does not allow the employment of computational methods for the solution of the optimization problem. However, this issue can be circumvented by introducing a terminal constraint for the state in the form of a Maximal Output Admissible Set (MAS) (Gilbert & Tan, 1991). This problem will be tackled in section 4. For now, it is sufficient to state that there exists a finite horizon within which enforcement of the constraints leads to enforcement of the constraints over an infinite horizon, given some reasonable assumptions on the plant dynamics (Rawlings & Muske, 1993).

3. Constraint relaxation approaches

3.1 Minimal-time approach

Minimal-time approaches allow constraint violations for a certain period of time, which is to be minimized. There is no commitment to reduce the peaks of the violations during this period. These are, respectively, the strongest advantage and the weakest drawback of these methods. The constraint violations are usually allowed to take place in the beginning of the control task, which reduces the time taken to achieve feasibility at the cost of degrading the transient response of the control-loop. Scokaert & Rawlings (1999) introduce an approach of minimal-time solution that considers the peak violation of the constraints as a secondary objective, after the minimization of the time to enforce the constraints. This avoids unnecessarily large peak violations.

One possibility to avoid control constraint violations, which are usually physical ones, is to enforce them while relaxing operating constraints on the state. This way, the problem always becomes feasible. One algorithm that implements a solution of this type may be stated as:

Data: $x(k)$
Result: Optimized control sequence $\hat{V}^*$
Solve constrained MPC problem;
if *infeasible* **then**
  Remove constraints on the state;
  Solve MPC problem;
  Find $\kappa = \kappa_{unc}$, which is the instant at which the state constraints are all enforced;
else
  Employ obtained control sequence;
  Terminate.
end
while *feasible* **do**
  $\kappa \leftarrow \kappa - 1$;
  Solve MPC problem with state constraints enforced from time κ until the end of the prediction horizon;
end
Employ last feasible control sequence;
Terminate.

Algorithm 1: Minimal-time algorithm

This algorithm determines the smallest time window over which the state constraints must be removed at the beginning of the prediction horizon in order to attain feasibility.

3.2 Soft-constraint approach

In this approach the cost function is modified to include a penalization on the violation of operating constraints. This way, a compromise is achieved between time and peak values of the violations, as well as performance of the control-loop. Scokaert & Rawlings (1999) propose the penalization of the sum of the square of the values of the violations instead of the peak as means to reduce their time length. This can be accomplished by simply adding slack variables to the state/output constraints of Eq. (7) in case of infeasibility and adding a term to the right-hand side of Eq. (8), as follows:

$$J_{Soft} = \sum_{i=0}^{N-1} \hat{v}^T(k+i|k)\Upsilon\hat{v}(k+i|k) + \epsilon_p^T W_{\epsilon_p}\epsilon_p + \epsilon_n^T W_{\epsilon_n}\epsilon_n \tag{21}$$

$$\begin{gathered} x_{P,min} - x_{ref} - \epsilon_n \leq x \leq x_{P,max} - x_{ref} + \epsilon_p, \\ \epsilon_p, \epsilon_n \geq 0 \end{gathered} \tag{22}$$

where W_{ϵ_n} and W_{ϵ_n} are positive-definite weight matrices. The additional restrictions $\epsilon_p, \epsilon_n \geq 0$ impose that the constraints are not made more restrictive than their original settings.

With the cost function of Eq. (21) subject to the constraints of Eq. (22), the amount by which each constraint is prioritized can be tuned by the choice of the weight matrices.

To this end, a rule of thumb known as "Bryson's rule" (Franklin et al., 2005), (Bryson & Ho, 1969) can be used as a guideline. It states that one may use the limits of the variables as parameters to choose their weights in the cost function so that their contribution is normalized. Therefore, the weights must be chosen so that the product between the admissible range (maximum value - minimum value) and the weight is approximately the same for all variables. However, in the present case, it is desirable that deviations of the slack variables from zero are more penalized than control deviations in order to enforce the constraints when possible. Therefore, it is reasonable to choose the weights for these variables an order of magnitude greater than the values obtained via Bryson's rule.

Scokaert & Rawlings (1999) discuss the inclusion of a linear term of penalization of the slack variables as means to obtain exact relaxations, i. e., the controller relaxes the constraints only when necessary. This can be achieved by tuning the weights of this term based on the Lagrange multipliers associated to the constrained minimization problem. However, an advantage of introducing terms that penalize the square of the slack variables is that the choice of a positive-definite weight matrix leads to a well-posed quadratic program, since the associated Hessian is positive definite.

3.3 Hard constraint relaxation with prioritization

There are methods which relax the operating constraints, possibly according to a priority list, in order to achieve feasibility of the optimization problem. There are various techniques employing such policies, some of which resort to optimization problems parallel to the MPC optimization in order to determine the minimum relaxation that is necessary to achieve feasibility. In this line, the priority list can be explored by solving many Linear Programming (LP) problems relaxing the constraints of lower priority until feasibility is achieved or by solving a single LP problem online as proposed by Vada et al. (2001). In their work, offline computations of the weights of the slack variables that relax the constraints are performed.

The calculated weights have the property of relaxing the constraints according to the defined priority in a single LP problem.

3.4 Simulation example

This example is based on a double integrator model, with sampling period of 1 time unit. Double integrators can be used to model a number of real-world systems, such as a vehicle moving in an environment where friction is negligible (space, for instance).

The discrete-time model matrices are:

$$A = \begin{bmatrix} 1 & 1 \\ 0 & 1 \end{bmatrix}, B = \begin{bmatrix} 0.5 \\ 1 \end{bmatrix} \tag{23}$$

and the LQR weight matrices are:

$$Q_{lqr} = \begin{bmatrix} 1 & 0 \\ 0 & 1 \end{bmatrix}, R_{lqr} = 1 \tag{24}$$

The control and prediction horizons were set to $M = 7$ and $N = 20$, respectively.

The constraints are: $-0.5 \leq x_1 \leq 0.5$ (position), $-0.1 \leq x_2 \leq 0.1$ (velocity) and $-0.01 \leq u \leq 0.01$ (acceleration).

A comparison between the results obtained with a minimal-time solution and a soft constraint approach is presented. Two choices of weight matrices were considered:

$$W^1_{\epsilon_n} = W^1_{\epsilon_p} = W^1 = \begin{bmatrix} 10 & 0 \\ 0 & 20 \end{bmatrix}, W^2_{\epsilon_n} = W^2_{\epsilon_p} = W^2 = \begin{bmatrix} 100 & 0 \\ 0 & 10000 \end{bmatrix} \tag{25}$$

The application of Bryson's rule to adjust the weight matrices would require the definition of an acceptable violation of the constraints, which could be established as the difference between physical and operating state constraints. However, since this example does not discriminate between these two types of constraints, the W^1 and W^2 matrices were chosen for the sole purpose of illustrating the effect of varying the weights.

The initial state of the system is $x_0 = [1.5 \;\; 0]^T$, which violates the constraints on x_1.

The first comparison involves the two infeasibility handling techniques (minimal-time and soft constraint). For this purpose, the W^1 weight matrix was employed. Figures 2 and 3 show the resulting state trajectories. It can be seen that the minimal-time approach leads to a faster recovery of feasibility, as the soft constraint approach takes longer to enforce all the constraints. This result can also be associated to the control profile presented in Fig. 4. In fact, the control obtained with the minimal-time approach reverses its sign earlier, as compared to the soft constraint approach.

The second comparison involves three scenarios: no state constraints and soft constraint approach with weights W^1 and W^2. Figures 5, 6 and 7 show the resulting state and control trajectories. As can be seen, a reduction in the weights tends to generate a solution closer to the unconstrained case. In fact, smaller weights on the slack variables result in a smaller penalization of the constraint violations. In the limit, if the weights are made equal to zero, the constraints can be relaxed as much as it is needed and therefore the unconstrained optimal solution is obtained.

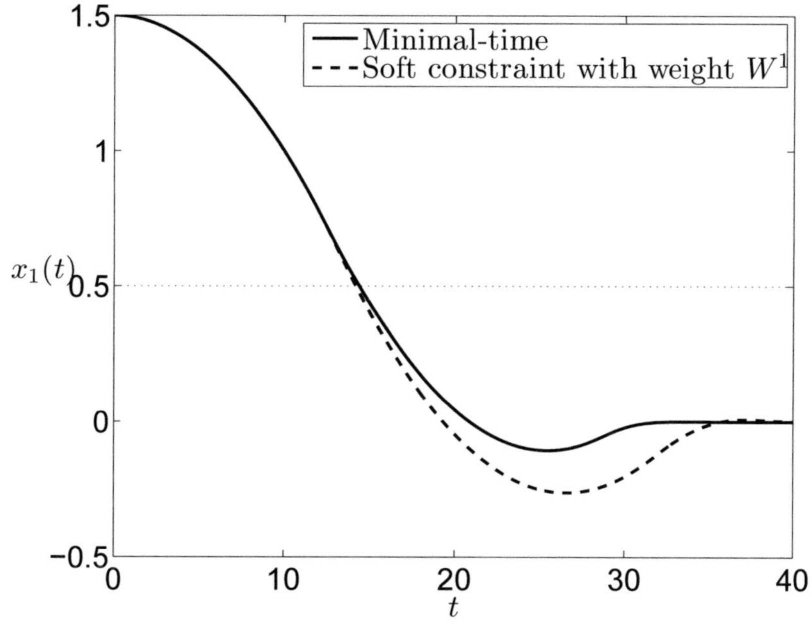

Fig. 2. Position (x_1) with constraint relaxation.

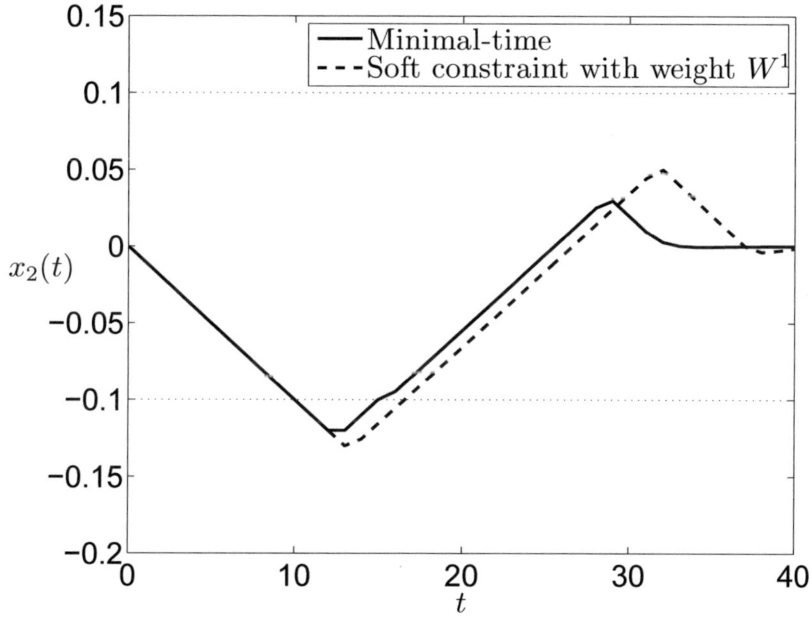

Fig. 3. Velocity (x_2) with constraint relaxation.

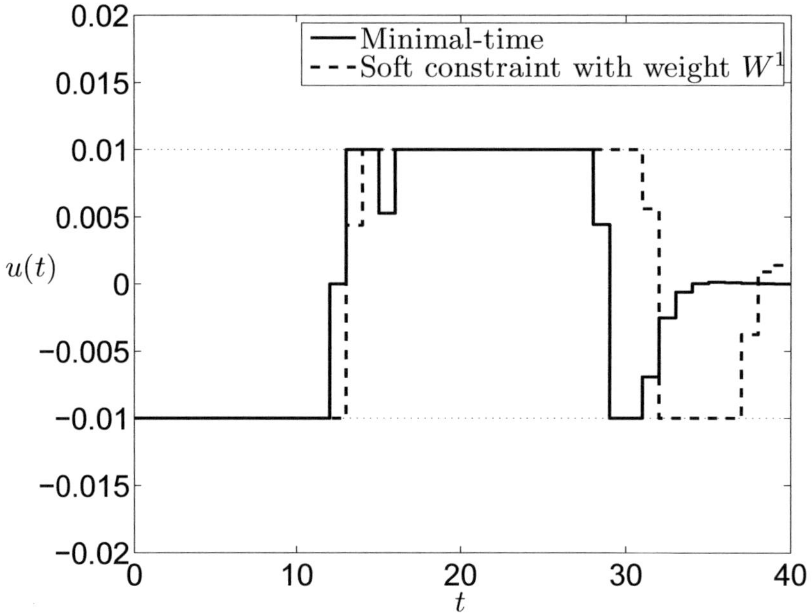

Fig. 4. Acceleration (u) with constraint relaxation.

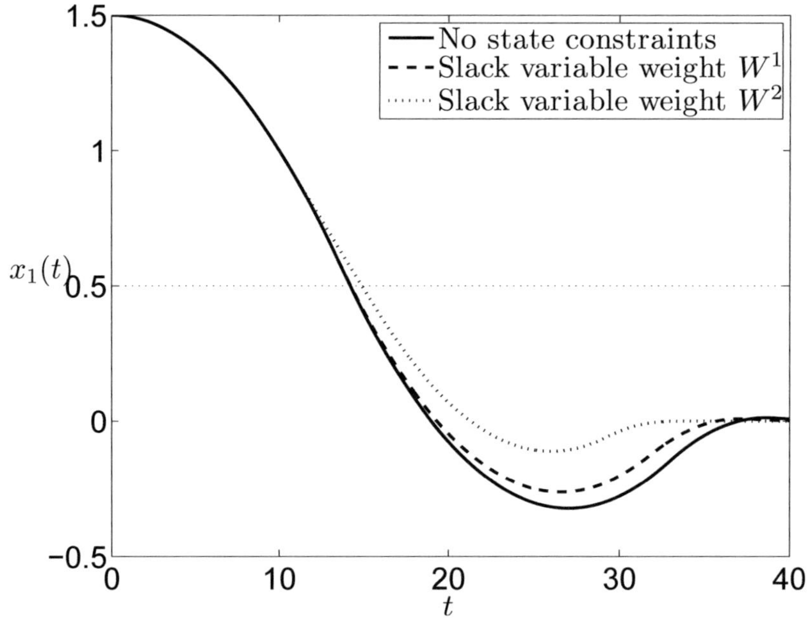

Fig. 5. Position (x_1) without state constraints and with soft constraint relaxation.

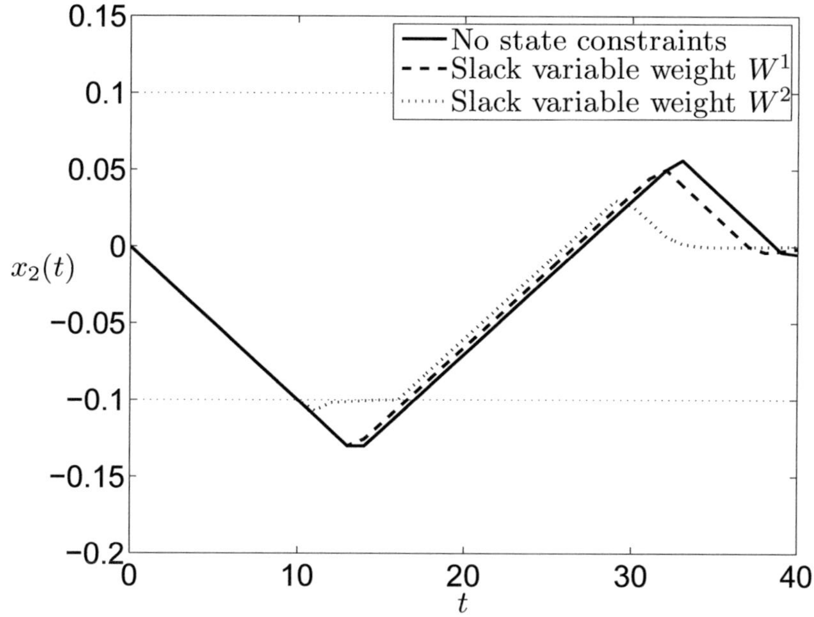

Fig. 6. Velocity (x_2) without state constraints and with soft constraint relaxation.

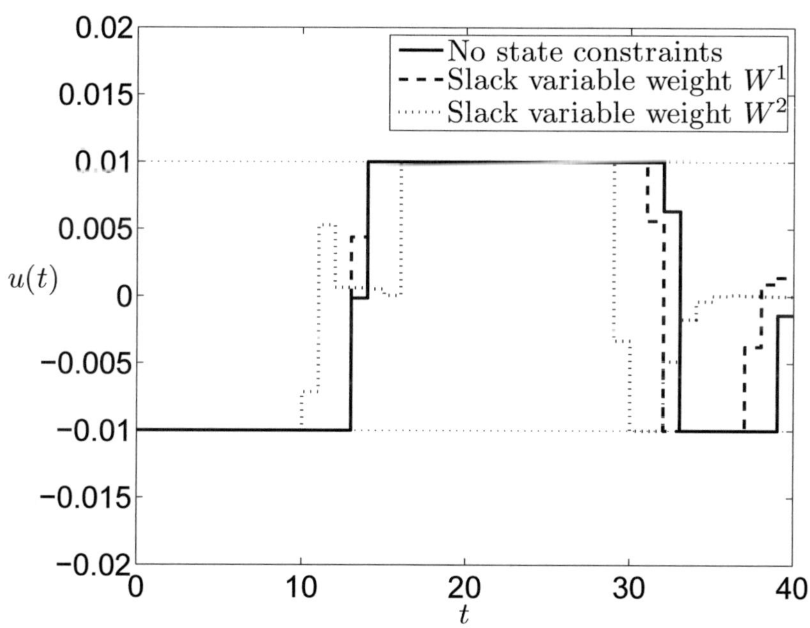

Fig. 7. Acceleration (u) without state constraints and with soft constraint relaxation.

4. Setpoint management approaches

The main idea behind setpoint management schemes is to find a new setpoint $x'_{ref}(k) = x_{ref}(k) - C\mu$ at each time k in order to make the problem feasible and to progressively steer the system state towards the original setpoint x_{ref}. $\mu \in \mathbb{R}^n$ is the setpoint management variable and $C \in \mathbb{R}^{q\times n}$ is a constant matrix. It is worth noting that, in the general case, changing the setpoint x_{ref} would also affect the corresponding setpoint u_{ref} for the control. As a result, the bounds on the control u would need to be changed, which would require the online recalculation of the terminal constraint set. Therefore, the class of systems considered in this study are restricted to those which require no adjustment in the control setpoint after a change in the state setpoint. This is a property of plants with integral behavior.

It is worth noting that these setpoint modifications impose a need of redetermination of the MAS every time the value of μ changes. The approach presented in the following subsection introduces a parameterization of the MAS in terms of the possible values of μ, avoiding the necessity to repeat the determination of the terminal set online.

4.1 Parameterization of the MAS

The parameterization of the MAS may be carried out through the employment of an augmented state vector $\bar{x}$ defined as (Almeida & Leissling, 2010)

$$\bar{x} = \begin{bmatrix} x \\ \mu \end{bmatrix}, \tag{26}$$

which evolves inside the MAS according to

$$\bar{x}(k+1) = \bar{A}\bar{x}(k), \quad \bar{A} = \begin{bmatrix} A - BK & 0 \\ 0 & I_n \end{bmatrix}. \tag{27}$$

It is worth noting that the identity matrix $I_n \in \mathbb{R}^{n\times n}$ multiplies the additional components of the state because these are supposed to remain constant along the prediction horizon. Although $\bar{A}$ has eigenvalues in the border of the unit circle (eigenvalues at $+1$ associated to the matrix I_n), it is still possible to determine the MAS in a finite number of steps because the dynamics given by Eq. (27) is stable in the Lyapunov sense (Gilbert & Tan, 1991).

The state constraints are altered by the management variable μ in the following fashion:

$$x_{P,min} - x_{ref} + C\mu \leq x \leq x_{P,max} - x_{ref} + C\mu \tag{28}$$

where C is a matrix that relates the vector $\mu \in \mathbb{R}^n$ of setpoint management variables to the corresponding component of the state vector $x \in \mathbb{R}^n$ whose setpoint is managed.

In order to incorporate the constraints to the parameterization, an auxiliary output variable $\bar{z}$ may be defined as

$$\bar{z} = \begin{bmatrix} x - C\mu \\ -x + C\mu \end{bmatrix} \tag{29}$$

which is subject to the following constraints:

$$\bar{z} \leq \begin{bmatrix} x_{P,max} - x_{ref} \\ x_{ref} - x_{P,min} \end{bmatrix} \tag{30}$$

Since $u = -Kx$ inside the MAS, the output function for the determination of the MAS becomes $\bar{z} = \bar{C}\bar{x}$ with

$$\bar{C} = \begin{bmatrix} I_n & -C \\ -I_n & C \end{bmatrix} \tag{31}$$

Having determined the MAS ($\bar{\mathbb{O}}_\infty$) associated to the dynamics of Eq. (27) with the constraints of Eq. (30), it can be particularized online by fixing the value of μ. The set $\bar{\mathbb{O}}_\infty$ obtained is invariant regarding matrix $\bar{A}$. It is convenient to note that the terminal constraint $\hat{x}(k+N|k) \in \bar{\mathbb{O}}_\infty$ for a particular choice of μ can replace the constraints from $i = N$ onwards in Eqs. (14) and (15). Imposing $\hat{x}(k+N|k) \in \bar{\mathbb{O}}_\infty$ is equivalent to imposing the constraints $\hat{u}(k+i|k) \in \mathbb{U}$ and $\hat{x}(k+i|k) \in \mathbb{X}$ until $i = N + t^*$, with t^* obtained during the offline determination of the parameterized MAS. Therefore, the infinite set of constraints of Eqs. (14) and (15) is reduced to a finite one.

4.2 Optimization problem formulation

Considering the setpoint management, the optimization problem to be solved at time k now involves $\hat{V}$ and μ as decision variables.

Thus, the optimization problem becomes

$$\min_{\hat{V},\mu} \hat{V}^T \Psi \hat{V} + \mu^T W_\mu \mu \tag{32}$$

s.t.

$$\begin{bmatrix} H_U \\ -H_U \\ H \\ -H \end{bmatrix} \hat{V} \leq \begin{bmatrix} \left[u_{max} - u_{ref}\right]_{N+t^*+1} - \Phi_U(x_P(k) - x_{ref} + C\mu) \\ \Phi_U(x_P(k) - x_{ref} + C\mu) - \left[u_{min} - u_{ref}+\right]_{N+t^*+1} \\ \left[x_{P,max} - x_{ref} + C\mu\right]_{N+t^*} - \Phi(x_P(k) - x_{ref} + C\mu) \\ \Phi(x_P(k) - x_{ref} + C\mu) - \left[x_{P,min} - x_{ref} + C\mu\right]_{N+t^*} \end{bmatrix}$$

where W_μ is a positive-definite weight matrix, the operator $[\bullet]_j$ stacks j copies of vector $\bullet$, and H, H_u, Φ and Φ_u are in accordance with Eq. 20.

The greater the weights in W_μ in comparison to Ψ, the closer the solution is to the one obtained without the need of setpoint management.

After the solution of the optimization problem of Eq. (32), the control signal to be applied to the plant is given by

$$u_P(k) = u_{ref} + \hat{v}^*(k|k) - K(x_P(k) - x_{ref} + C\mu^*) \tag{33}$$

4.3 Simulation example

The simulation scenario employed in this example is the same as that of subsection 3.4. Only the constraints over the position variable are different ($-1 \leq x_1 \leq 1$). The determination of the MAS leads to $t^* = 7$ and M remains equal to 7. Therefore, the constraint horizon in order to guarantee that the constraints are enforced over an infinite horizon is $N = M + t^* = 14$.

The initial state is $x_0 = [1 \;\; 0]^T$, which respects the constraints. However, the problem is infeasible, making the employment of a technique to recover feasibility mandatory. The procedure described in this section can be used to recover feasibility. The setpoint of the position is chosen for management, meaning that $\mu \in \mathbb{R}$ and

$$C = \begin{bmatrix} 1 \\ 0 \end{bmatrix} \tag{34}$$

It is desirable to keep the setpoint management as close to zero as possible. To this end, the weight of the setpoint management variable is chosen as $W_\mu = 1000$.

Figure 8 shows the position variable, which starts at the edge of the constraint and is steered to the origin without violating the constraints.

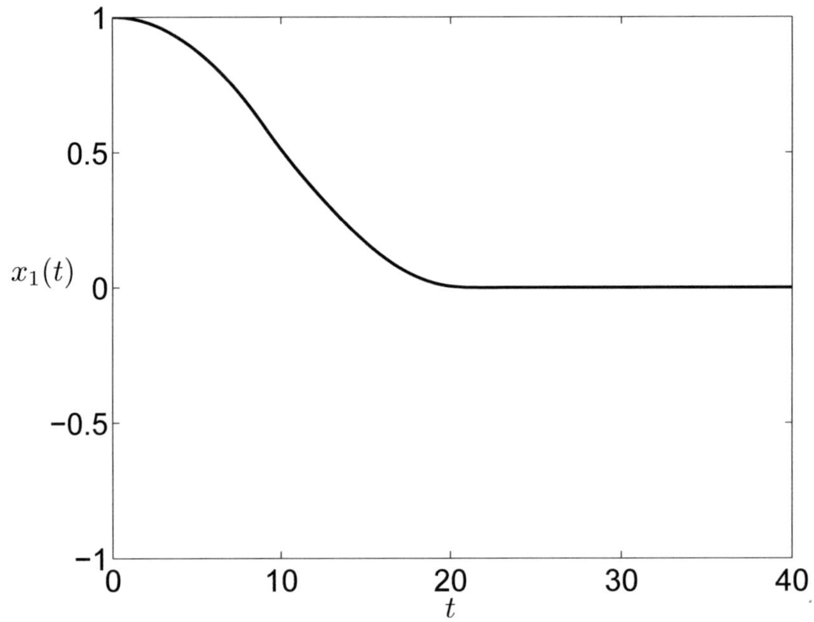

Fig. 8. Position (x_1) with setpoint management.

It can be seen in Fig. 9 that the velocity variable gets close to its lower bound (-0.1), but this constraint is also satisfied. Figure 10 shows that the constraints on the acceleration are active in the beginning of the maneuver, but are not violated.

The setpoint management variable μ is shown in Fig. 11. It can be seen that the management technique is applied up to time $t = 10$. This time coincides with the change in the acceleration from negative to positive.

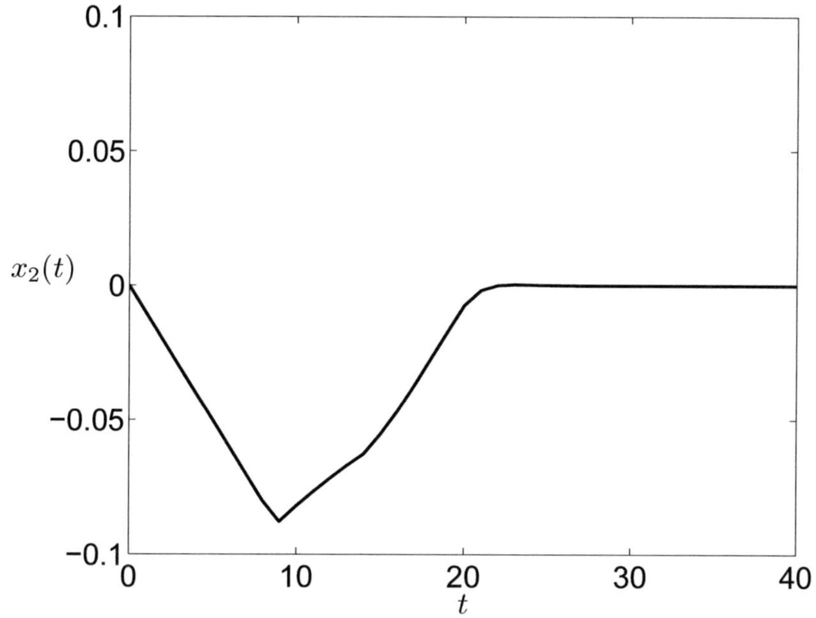

Fig. 9. Velocity (x_2) with setpoint management.

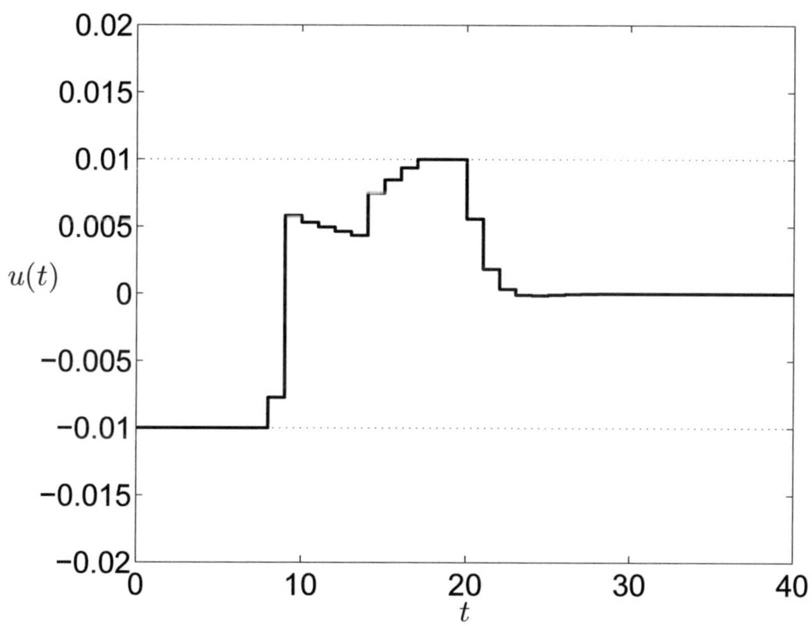

Fig. 10. Acceleration (u) with setpoint management.

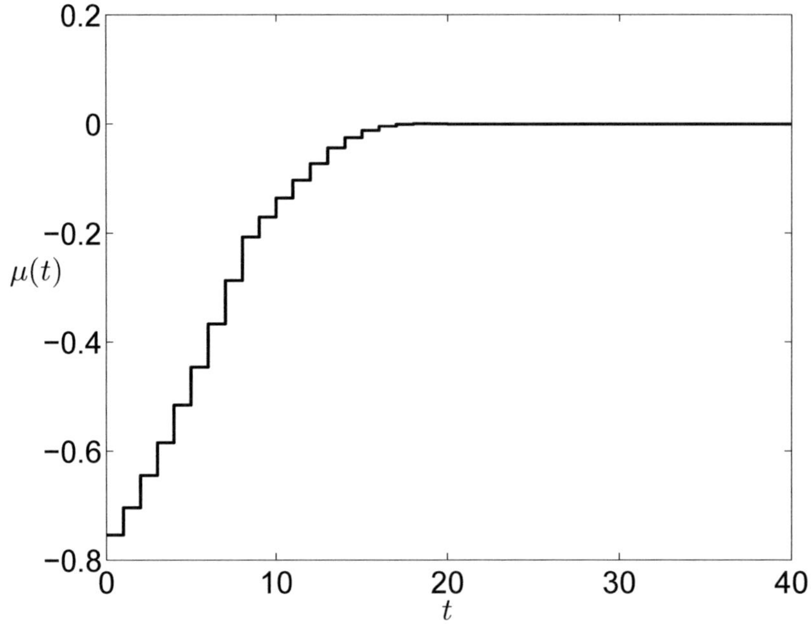

Fig. 11. Position setpoint management variable (μ).

5. Conclusions

In real applications of MPC controllers, noise, disturbances, model-plant mismatches and faults are commonly found. Therefore, infeasibility of the associated optimization problem can be a recurrent issue. This justifies the study of techniques capable of driving the system to a feasible region, since infeasibility may cause prediction errors, deployment of impracticable control sequences and instability of the control loop. Computational workload is also of great concern in real applications, thus the adopted techniques must be simple enough to be executed in a commercial off-the-shelf computer within the sample period and effective enough to make the problem feasible. In this chapter a review of the literature regarding feasibility issues was presented and two of the more widely adopted approaches (constraint relaxation and setpoint management) were described. Simulation examples of some illustrative techniques were presented in order to clarify the advantages, drawbacks and difficulties in implementation of some techniques.

6. Acknowledgements

The authors acknowledge the financial support of FAPESP (MSc scholarship 2009/12674-0) and CNPq (research fellowship).

7. References

Afonso, R. J. M. & Galvão, R. K. H. (2010a). Controle preditivo com garantia de estabilidade nominal aplicado a um helicóptero com três graus de liberdade empregando relaxamento de restrições de saída (Predictive control with nominal stability guarantee applied to a helicopter with three degrees of freedom employing output constraint relaxation - text in portuguese), *Proc. XVIII Congresso Brasileiro de Automática*, pp. 1797 – 1804.

Afonso, R. J. M. & Galvão, R. K. H. (2010b). Predictive control of a helicopter model with tolerance to actuator faults, *Proc. Conf. Control and Fault-Tolerant Systems (SysTol), 2010*, pp. 744 – 751.

Almeida, F. A. & Leissling, D. (2010). Fault-tolerant model predictive control with flight-test results, *J. Guid. Control Dyn.* 33(2): 363 – 375.

Alvarez, T. & de Prada, C. (1997). Handling infeasibilities in predictive control, *Computers & chemical engineering* 21: S577 – S582.

Bemporad, A., Casavola, A. & Mosca, E. (1997). Nonlinear control of constrained linear systems via predictive reference management, *IEEE Trans. Automatic Control* 42(3): 340 – 349.

Bemporad, A. & Mosca, E. (1994). Constraint fulfilment in feedback control via predictive reference management, *Proc. 3rd IEEE Conf. Control Applications*, Glasgow, UK, pp. 1909 – 1914.

Bryson, A. E. & Ho, Y.-C. (1969). *Applied Optimal Control*, Blaisdell, Waltham, MA.

Chisci, L., Rossiter, J. A. & Zappa, G. (2001). Systems with persistent disturbances: predictive control with restricted constraints, *Automatica* 37(7): 1019–1028.

Franklin, G., Powell, J. & Emami-Naeini, A. (2005). *Feedback Control of Dynamic Systems*, 5th edn, Prentice Hall, Upper Saddle River, NJ.

Gilbert, E. G. & Kolmanovsky, I. (1995). Discrete-time reference governors for systems with state and control constraints and disturbance inputs, *Proc. 34th IEEE Conference on Decision and Control*.

Gilbert, E. G. & Tan, K. T. (1991). Linear systems with state and control constraints: the theory and application of maximal output admissible sets, *IEEE Trans. Automatic Control* 36(9): 1008–1020.

Kapasouris, P., Athans, M. & Stein, G. (1988). Design of feedback control systems for stable plants with saturating actuators, *Proc. 27th IEEE Conference on Decision and Control*.

Kouvaritakis, B., Rossiter, J. A. & Cannon, M. (1998). Linear quadratic feasible predictive control, *Automatica* 34(12): 1583–1592.

Limon, D., Alvarado, I., Alamo, T. & Camacho, E. (2008). MPC for tracking piecewise constant references for constrained linear systems, *Automatica* 44(9): 2382–2387.

Maciejowski, J. M. (2002). *Predictive Control with Constraints*, 1st edn, Prentice Hall, Harlow, England.

Montandon, A. G., Borges, R. M. & Henrique, H. M. (2008). Experimental application of a neural constrained model predictive controller based on reference system, *Latin American applied research* 38: 51 – 62.

Rawlings, J. & Muske, K. (1993). The stability of constrained receding horizon control, *IEEE Trans. Automatic Control* 38(10): 1512–1516.

Rodrigues, M. A. & Odloak, D. (2005). Robust mpc for systems with output feedback and input saturation, *Journal of Process Control* 15: 837 – 846.

Rossiter, J. A. (2003). *Model-based Predictive Control: a practical approach*, 1st edn, CRC Press, Boca Raton.

Scokaert, P. (1994). *Constrained Predictive Control*, PhD thesis, Univ. Oxford, UK.

Scokaert, P. & Rawlings, J. (1998). Constrained linear quadratic regulation, *IEEE Trans. Automatic Control* 43(8): 1163–1169.

Scokaert, P. & Rawlings, J. (1999). Feasibility issues in linear model preditctive control, *AIChE Jounal* 45(8): 1649 – 1659.

Vada, J., Slupphaug, O., Johansen, T. & Foss, B. (2001). Linear mpc with optimal prioritized infeasibility handling: application, computational issues and stability, *Automatica* 37(11): 1835 – 1843.

Zafiriou, E. & Chiou, H. (1993). Output constraint softening for siso model predictive control, *American Control Conference*.

Part 2

Recent Applications of MPC

4

Predictive Control Applied to Networked Control Systems

Xunhe Yin[1,2], Shunli Zhao[1], Qingquan Cui[1,3] and Hong Zhang[4]
[1]*School of Electric and Information Engineering, Beijing Jiaotong University,*
[2]*School of Electrical and Information Engineering,*
University of Sydney, Sydney,
[3]*Yunnan Land and Resources Vocational College,*
Kunming,
[4]*Beijing Municipal Engineering Professional Design*
Institute Co.Ltd, Beijing,
[1,3,4]*China*
[2]*Australia*

1. Introduction

The researches of the networked control systems (NCSs) cover a broader, more complex technology, because that networked control systems relate to computer network, communication, control, and other interdisciplinary fields. Networked control systems have become one of the hot spots of international control areas in recent years. The networked control system theoretical research is far behind its application, so the networked control system theory study has important academic value and economic benefits at present.

NCSs performance is not only related with the control algorithms, but also the network environment and the scheduling algorithms. The purpose of network scheduling is to avoid network conflicts and congestion, accordingly reducing the network-induced delay, packet loss rate and so on, which can ensure the better network environment. If the case, where the data cannot be scheduled, appears in the network, the control algorithm has not fundamentally improved the performance of the system, thus only adjusting data transmission priorities and instants over the network by using the scheduling algorithms, in order to make the whole system to achieve the desired performance.

Along with the networked control system further research, people gradually realized that the scheduling performance must be taken into account when they research control algorithms, that is, considering the two aspects of scheduling and control synthetically. The joint design of both scheduling performance and control performance is concerned by the majority of researchers (Gaid M B et al., 2006a,2006b; Arzen K E et al., 2000). Therefore, NCSs resource scheduling algorithms, as well as scheduling and control co-design are the main research directions and research focus.

The generalized predictive control and the EDF (Earliest Deadline First) scheduling algorithm are adopted by the NCSs co-design in this chapter. The co-design method considers both the NCSs scheduling performance and control performance, and then the process of the general co-design method is also given. From the TrueTime simulation results based on NCSs with three loops of DC-motors, NCSs under co-design compared with NCSs without co-design, we can find that the former shows better control performance and scheduling performance, and a better anti-jamming ability and adaptive ability for network, so that the NCSs with co-design can guarantee to operate in an optimal state.

2. Brief review of Generalized Predictive Control

GPC (Generalized Predictive Control) algorithm is proposed by Clarke et al (Calrke & Mohtadi, 1989) in the 80s of last century, as a new class of predictive control algorithm. The algorithm is based on Controlled Auto-Regressive Integrated Moving Average (CARIMA) model, adopts an optimization of the long time indicators combined with the identification and self-correcting mechanism, shows strong robustness and has broad scope of application. The significance of GPC algorithm is that the algorithm can still get sub-optimal solution when mismatch or time-varying occurs in the controlled plant model, so it has strong robustness, but also can eliminate the static error of the system with using CARIMA model., The generalized predictive control, which is optimized control algorithms based on the prediction model, rolling optimization and online feedback correction, have distinct characteristics as a new type of control algorithms. (Wang et al., 1998; Guan & Zhou, 2008; Ding, 2008).

2.1 Prediction model

Refer to the generalized predictive control; the controlled plant is usually represented by the model of CARIMA:

$$Ay(k) = Bu(k-1) + C\frac{\xi(k)}{\Delta} \tag{1}$$

where $u(k)$ and $y(k)$ are control input and system output respectively, $\xi(k)$ is a white noise with zero mean and standard deviation σ^2, $\Delta = 1 - z^{-1}$ is a difference operator, $A = 1 + a_1 z^{-1} + \cdots + a_n z^{-n}$, $B = b_1 z^{-1} + \cdots + b_n z^{-n}$, $C = 1 + c_1 z^{-1} + \cdots + c_n z^{-n}$.

To simplify the inference process of the principle, without loss of generality, let C=1. To derive the optimization prediction value of $y(k+j)$ after j steps, the Diophantine equation is considered firstly:

$$I = E_j(z^{-1})A(z^{-1})\Delta + z^{-j}F_j(z^{-1}) \tag{2}$$

where $E_j(z^{-1}) = e_{j,0} + e_{j,1}z^{-1} + \cdots + e_{j,j-1}z^{-j+1}$, $F_j(z^{-1}) = f_{j,0} + f_{j,1}z^{-1} + \cdots + f_{j,n}z^{-n}$, they are multinomial which are decided by the model parameter A and prediction length j, $e_{j,0} \cdots e_{j,j-1}$ and $f_{j,0} \cdots f_{j,j-1}$ are coefficients.

Using $E_j \Delta z^j$ to multiply both sides of (1), then combining (2), $y(k+j)$ is derived:

$$y(k+j) = E_j B \Delta u(k+j-1) + F_j y(k) + E_j \xi(k+j) \tag{3}$$

By the expressions E_j, can see that $E_j\ \xi\ (k+j)$ is an unknown noise starting from instant k^{th}, the output prediction value of the futurity j steps starting from instant k^{th} are derived after deleting the term $E_j \xi(k+j)$:

$$\hat{y}(k+j) = E_j B\Delta u(k+j-1) + F_j y(k) \tag{4}$$

Let $G_j = E_j B$, and $j = 1,2\cdots,N$, (4) can be written as matrix equation (5):

$$\hat{\boldsymbol{y}} = \boldsymbol{G}\boldsymbol{\Delta u} + \boldsymbol{f} \tag{5}$$

where $\hat{\boldsymbol{y}} = [y(k+1)\ y(k+2)\ ...y(k+N)]$, $\boldsymbol{\Delta u} = [\Delta u(k)\ \Delta u(k+1)\ ...\Delta u(k+M-1)]$, N is the model time domain while M is the control time domain, $\boldsymbol{f} = [f_1(k)\ f_2(k)\ ...f_N(k)]^T$, $f_j(k) = z^{n-1}[\boldsymbol{G}_n - z^{1-n}g_{n,n-1} - \cdots - g_{n,0}]\Delta u(k) + F_j y(k), n = 1,2,\cdots N$,

$$\boldsymbol{G} = \begin{bmatrix} g_1 & 0 & \cdots & 0 \\ g_2 & g_1 & \cdots & 0 \\ \vdots & \vdots & \ddots & \vdots \\ g_p & g_{p-1} & \cdots & g_1 \\ \vdots & \vdots & \ddots & \vdots \\ g_N & g_{N-1} & \cdots & g_{N-M+1} \end{bmatrix}_{N\times M}$$

2.2 Rolling optimization

To enhance the robustness of the system, the quadratic performance index with output error and control increment weighting factors are adopted:

$$J = \sum_{j=N_0}^{P} [y(k+j) - y_r(k+j)]^2 + \sum_{j=1}^{M} [\lambda(j)\Delta u(k+j-1)]^2 \tag{6}$$

where N_0 is the minimum prediction horizon, and $N_0 \geq 1$, P is the maximum prediction horizon, M is the control horizon, that means the control value will not be changed after M steps, $\lambda(j)$, which is a constant λ in the general control systems, is the control increment weighting factor, but it will be adjusted in real time within the control process in the co-design of control and scheduling to ensure optimal control.

The optimal control law is as follow:

$$\boldsymbol{\Delta u}(k) = (\boldsymbol{G}^T\boldsymbol{G} + \lambda \boldsymbol{I})^{-1}\boldsymbol{G}^T[\boldsymbol{y}_r(k+1) - \boldsymbol{f}] \tag{7}$$

Then the incremental series of open loop control from instant k^{th} to instant ($k+M-1)^{th}$ is derived after expanding the formula (7):

$$\boldsymbol{\Delta u}(k+i-1) = \boldsymbol{d}_i^T[\boldsymbol{y}_r(k+1) - \boldsymbol{f}] \tag{8}$$

where $\boldsymbol{d}_i^T$ is the ith increment of $(\boldsymbol{G}^T\boldsymbol{G} + \lambda \boldsymbol{I})^{-1}\boldsymbol{G}^T$, $\boldsymbol{d}_i^T = [d_{i1}\ d_{i1}\ \cdots d_{iP}]$.

In the real control systems, the first control variable will be used in every period. If the control increment $\boldsymbol{\Delta u}(k)$ of the current instant k^{th} is executed, the control increment after k^{th} will be recalculated in every period, that is equivalent to achieve a closed loop control strategy, then the first raw of $(\boldsymbol{G}^T\boldsymbol{G} + \lambda \boldsymbol{I})^{-1}\boldsymbol{G}^T$ is only necessary to recalculate. So the actual control action is denoted as (9):

$$u(k) = u(k-1) + d_l^{T}[y_r(k+1) - f] \tag{9}$$

2.3 Feedback correction

To overcome the random disturbance, model error and slow time-varying effects, GPC maintains the principle of self-correction which is called the generalized correction, by constantly measuring the actual input and output, estimates the prediction model parameters on-line. Then the control law is corrected.

The plant model can be written as:

$$A\Delta y(k) = B\Delta u(k-1) + \xi(k)$$

Then we can attain $$\Delta y(k) = -(A-1)\Delta y(k) + B\Delta u(k-1) + \xi(k) \tag{10}$$

Model parameters and data parameters are expressed using vector respectively

$$\theta = [a_1 \cdots a_n \vdots b_0 \cdots b_m] \tag{11}$$

$$\varphi = [-\Delta y(k-1) \cdots -\Delta y(k-n) \vdots -\Delta u(k-1) \cdots -\Delta u(k-m+1)] \tag{12}$$

Then the above equation (10) can be written into the following form:

$$\Delta y(k) = \varphi^{T}(k)\theta + \xi(k) \tag{13}$$

The model parameters can be estimated by recursive least squares method with forgetting factor. The parameters of polynomial A, B are obtained by identification. d_i^T and f in control law of equation (9) can be recalculated, and that the optimal control $u(k)$ is found.

2.4 Generalized predictive control performance parameters

Generalized predictive control performance parameters (Ding, 2008; Li, 2009) contain minimum prediction horizon N_0, maximum prediction horizon *P*, control horizon *M*, and control weighting factor λ.

1. Minimum prediction horizon N_0

When the plant delay *d* is known, then take $N_0 \geq d$. If $N_0 \geq d$, there are some output of $y(k+1),\cdots,\ y(k+P)$ without the impact from input $u(k)$, this will waste some computation time. When *d* is unknown or varying, generally let N_0 =1, that means the delay may be included in the polynomial $B(z^{-1})$.

2. Maximum prediction horizon *P*

In order to make the rolling optimization meaningfully, *P* should include the actual dynamical part of the plant. Generally to take *P* close to the rise time of the system, or to take *P* greater than the order of $B(z^{-1})$. In practice, it is recommended to use a larger *P*, and make it more than the delay part of the impulse response of the plant or the reverse part caused by the non-minimum phase, and covers the main dynamic response of the plant. The size of P has a great effect on the stability and rapidity of the system. If P is small, the dynamic performance is good, but with poor stability and robustness. If P is big, the robustness is good, but the dynamic performance is bad, so that system's real-time performance is reduced because of increasing of computing time. In the actual application, we can choose the one between the two values previously mentioned to make the closed-loop system not only with the desired robustness but also the required dynamic performance (rapidity) (Ding, 2008).

3. Control horizon M

This is an important parameter. Must $M \leq P$, because that the optimal prediction output is affected by P control increment values at best. Generally, the M is smaller, the tracking performance is worse. To improve the tracking performance, increasing the control steps to improve the control ability for the system, but with the increase of M, the control sensitivity is improved while the stability and robustness is degraded. And when M increases, the dimension of the matrix and the calculation amount is increased; the real-time performance of the system is decreased, so M should be selected taking into account the rapidity and stability.

4. Control weighting factor λ

The effect of the control weighting factor is to limit the drastic change of the control increment, to reduce the large fluctuation to the controlled plant. The control stability is achieved by increasing λ while the control action is weakened (Li, 2009). To select small number λ generally, firstly let λ is 0 or a smaller number in practice. If the control system is steady but the control increment changes drastically, then can increase λ appropriately until the satisfactory control result is achieved.

3. EDF scheduling algorithm and network performance parameters

3.1 EDF scheduling algorithm

EDF scheduling algorithm is based on the length of the task assigned from deadline for the priority of the task: the task is nearer from the required deadline and will obtain the higher priority. EDF scheduling algorithm is a dynamic scheduling algorithm, the priority of the task is not fixed, but changes over time; that is, the priority of the task is uncertain. EDF scheduling algorithm also has the following advantages except the advantages of the general dynamic scheduling algorithm:

1. can effectively utilize the network bandwidth resources, and improve bandwidth utilization;
2. can effectively analyze schedulability of information that will be scheduled;
3. is relatively simple to achieve it, and the executed instructions is lessr in the nodes.

For N mutual independent real-time periodic tasks, when the EDF algorithm is used, the schedulability condition is that the total utilization of the tasks meets the following inequality:

$$U = \sum_{i=1}^{N} \frac{c_i}{T_i} \leq 1 \tag{14}$$

where c_i is the task execution time, T_i is the task period. In NCSs, c_i is the data packet the sampling time, T_i is the data sampling period.

EDF scheduling algorithm can achieve high utilization from the point of resource utilization, and meet the conditions for more information needs under the same condition of resource, thus it will increase the utilization of resources. Furthermore, EDF is a dynamic scheduling algorithm, and it can dynamically adjust the priority of the message, and lets the limited resources make a more rational allocation under the case of heavy load of information, and makes some soft real-time scheduling system can achieve the desired performance under the condition of non-scheduling.

Suppose there are two concurrent real-time periodic tasks need to be addressed, the execution time of the two messages is 5ms, and the sampling periods are 8ms and 10ms respectively, and suppose the deadline for all information equal to their sampling period. The total utilization of the information is:

$$U = \frac{5}{8} + \frac{5}{10} = 1.125 > 1$$

By the schedulability conditions (14) of EDF, we know that EDF scheduling algorithm is not scheduled; in this case, co-design of scheduling and control is potential to research and solve this type of problem.

3.2 Network performance parameters

Network performance parameters include: network-induced delay, network bandwidth, network utilization, packet transmission time. The EDF scheduling algorithm is also related to the sampling period, priority, and deadline. The greater the network-induced delay is, the poorer is the network environment; data transmission queue and the latency are longer, whereas the contrary is the shorter. The network bandwidth is that the amount of information flows from one end to the other within the specified time, is the same as the data transfer rate, and network bandwidth is an important indicator for the measure of network usage. The network bandwidth is limited in a general way. When the data transmitted per unit time is greater than the amount of information of network bandwidth, network congestion will occur and network-induced delay is larger, thus impacting on the data in real time. The sampling period is an important parameter of network scheduling, but also associated to control performance of the system; the specific content will be described in the next section.

4. Co-design optimization method

4.1 Relationship between sampling period and control performances

In networked control system, which is a special class of digital control system, the feedback signal received by the controller is still periodic sampling data obtained from sensor, but these data transmitted over the network, rather than the point to point connection. The network can only be occupied by a task in certain instant, because that network resources are shared by multiple tasks; in other words, when one task is over the network, the other ones will wait until the network is free. In this case, the feedback signal sampling period and the required instant of feedback signal over network will jointly determine control system performance.

Although the controller requires sampling period as small as possible for getting feedback signal more timely, the smaller sampling period means the more times frequently need to send data in network, so that the conflict occurs easily between tasks, data transmission time will increase in the network, and even the loss of data may occur.

However, sampling period cannot too large in the network, because that larger sampling period can decrease the transmission time of the feedback signal in the network, but will not fully utilize network resources. Therefore, the appropriate sampling period must be selected in the practical design in order to meet both the control requirements and the data transmission stability in the network, and finding the best tradeoff point of sampling period to use of network resources as full as possible, thereby enhancing the control system performance (Li, 2009).

Fig.1 shows the relationship between the sampling period and control performance (Li et al., 2001), it clearly illustrates the effect of sampling period on continuous control system, digital control system and networked control system, the meanings of T_A, T_B and T_C are also defined.

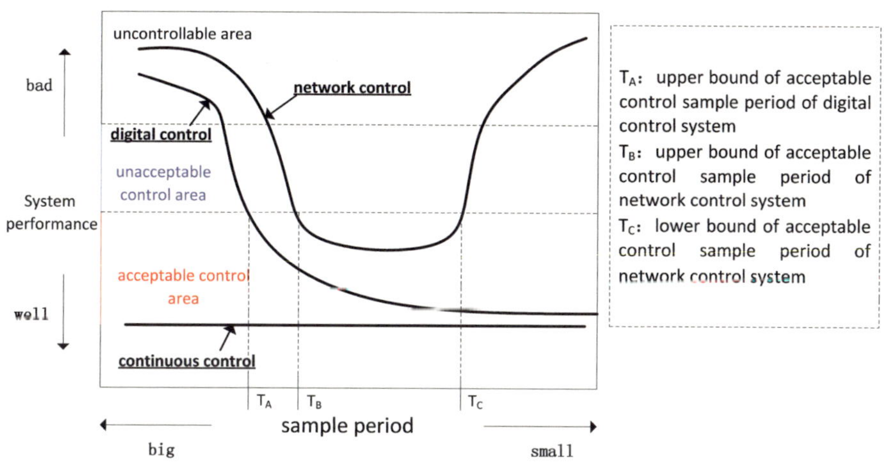

Fig. 1. The impact of Sampling period on control system performance

By analyzing the impact of sampling period for the control system performance, we see that changing the sampling period is very important to the networked control system performance. According to the different requirements for loops of NCSs, it has great significance for improving the system performance by changing the network utility rate of each loop and further changing the sampling period of each loop.

4.2 Joint optimization of the sampling periods

In NCSs, sampling period has effect on both control and scheduling, the selection of sampling period in NCSs is different from the general computer control system. Considering both the control performance and network scheduling performance indicators to optimize the sampling period of NCSs is the main way to achieve the co-design of control and scheduling (Zhang & Kong, 2008).

In NCSs, in order to ensure the control performance of the plant, generally the smaller sampling period is needed, but the decreased sampling period can lead the increased transmission frequency of the packets, and increase the burden of the network scheduling, therefore, control and scheduling are contradictory for the requirements of sampling period. The sampling periods of sensors on each network node not only bound by the stability of the plant but also the network schedulability. The way to solve this problem is to compromise the control performance and scheduling performance under certain of constraint conditions, and then to achieve the overall optimal performance of NCSs (Guan & Zhou, 2008; Zhang & Kong, 2008).

1. The selection of the objective function

Sampling period is too large or too small can cause deterioration of the system output performance, therefore, to determine the optimal sampling period is very important for the co-design of control and scheduling in NCSs. From the perspective of control performance, the smaller the sampling period of NCSs is, the better is its performance; from the perspective of scheduling performance, it will have to limit the decrease of the sampling period due to network communication bandwidth limitations. Optimization problem of the sampling period can be attributed to obtain the minimum summation of each control loop performance index function (objective function) under the conditions that the network is scheduling and the system is stable.

Suppose the networked control system optimal objective function is $J_{\min}$, then

$$J_{\min} = \sum_{i=1}^{N} p_i J_i \tag{15}$$

where p_i is weight, the greater the priority weight value of the network system is, the more priority is the data transmission . J_i is the performance index function of loop i, N is the total number of control loops.

2. Scheduling constraints

In order to make control information of networked control system transmit over the network effectively, meet the real-time requirements of period and control tasks, network resources allocation and scheduling are necessary. It ensures the information of control tasks to complete the transfer within a certain period of time to ensure the timeliness of the data and improve the network utilization. In this chapter, single packet transmission of information is analyzed, and the scheduling is non-priority.

Different scheduling algorithms correspond to the different schedulability and sampling period constraints. Currently, the commonly used network scheduling algorithms are: static scheduling algorithm, dynamic scheduling, mixed scheduling algorithm, and so on.

For static scheduling algorithm, such as *RM* algorithm, the following scheduling constraints can be chosen (Guan & Zhou, 2008):

$$\frac{c_1}{T_1}+\frac{c_2}{T_2}+\ldots+\frac{c_i}{T_i}+\frac{\bar{b}_{l,i}}{T_i}\le i(2^{\frac{1}{i}}-1) \tag{16}$$

where T_i, c_i and $\bar{b}_{l,i}$ are the sampling period, transmission time and congestion time of ith control loop respectively. $\bar{b}_{l,i}=\max\limits_{j=i+1,\ldots,N} c_j$ is the congestion time of the worst time which means the current task is blocked by the low priority task.

For dynamic scheduling, such as EDF algorithm, the following scheduling constraints can be chosen (Pedreiras P & Almenida L, 2002):

$$U=\sum_{i=1}^{N}\frac{c_i}{T_i}\le 1 \tag{17}$$

T_i, c_i are the sampling period and the data packet transmission time of ith control loop respectively.

3. Stability conditions of the system

The upper limit of the sampling period of networked control systems with delay (Mayne et al.,2003) is:

$$\boldsymbol{T}_{max}=\frac{\boldsymbol{T}_{bw}}{20}-2\tau_i \tag{18}$$

where T_{max} is the maximum value of the sampling period, ω_{bw} is the system bandwidth, T_{bw} is derived by ω_{bw}, τ_i is the network induce delay of loop i.

EDF scheduling algorithm is used in this chapter, the optimization process of the compromised sampling period of overall performance of the NCSs can be viewed as an optimization problem.

Objective function :

$$J_{\min}=\sum_{i=1}^{N}p_i J_i$$

Constraint condition:

$$\boldsymbol{T}_{max}=\frac{\boldsymbol{T}_{bw}}{20}-2\tau_i$$

$$U=\sum_{i=1}^{N}\frac{c_i}{T_i}\le 1$$

The constraints of network performance and control performance are added in the problem above simultaneously. They ensure the system to run on a good performance under a certain extent.

However, the optimal design method takes into account the relatively simple elements of the networked control system, and the involved performance parameters are less. So adding more network scheduling parameters and system control parameters is necessary to optimize the design jointly. An optimization method of taking both scheduling performance and control performance is proposed for system optimization operation. The core idea of the proposed methods is to make the interaction between the two performance indicators of networked control system---network scheduling performance and control performance, which affect on the system stable and efficient operation, so as to ensure network performance and control performance in NCSs.

4.3 Joint optimization of predictive control parameters

The preferences of GPC can be considered from two aspects. For general process control, let N_0=1, P is the rise time of the plant, M =1, then the better control performance is achieved. For the higher performance requirements of the plant, such as the plant in NCS, needs a bigger P based on the actual environment. A large number of computer simulation studies (Mayne et al., 2003; Hu et al., 2000; Chen et al., 2003)have shown that P and λ are the two important parameters affecting GPC control performance. When P increases, the same as λ, the smaller λ and the bigger P will affect the stability of the close loop system. The increase of the two parameters λ and P will slow down the system response speed, on the contrary, P less than a certain value will result in the system overshoot and oscillation.

When network induce delay $\tau_i < T$ (T is the sampling period), based on the above analysis of control and network parameters affecting on NCSs performance, network environment parameters will be considered in the follows: network induce delay, network utilization and data packet transmission time. The optimal rules of prediction control parameters are determined by the following three equations of loop i:

$$M_i(k+1) = M_i(k) + [(\frac{\Delta\tau_i}{\tau_i} + \frac{\Delta U}{U})\omega_1] \tag{19a}$$

$$P_i(k+1) = P_i(k) + [(\frac{\Delta U}{U} - \frac{\Delta\tau_i}{\tau_i} + \frac{\Delta c_i}{c_i})\omega_2] \tag{19b}$$

$$\lambda_i(k+1) = \lambda_i(k) + [(\frac{\Delta\tau_i}{\tau_i} + \frac{\Delta U}{U})\omega_3] \tag{19c}$$

where $M_i(k)$ is the control domain of loop i at sampling instant k^{th}, $P_i(k)$ is the minimum prediction domain of loop i at sampling instant k^{th}, $\lambda_i(k)$ is the control coefficient of loop i at sampling instant k^{th}, $\{\omega_1, \omega_2, \omega_3\}$ is the quantization weight, U is the network utilization, τ_i is the network induce delay of loop i, c_i is the data transmission time of loop i, $\Delta\tau_i$ is the error change of network induce delay, Δc_i is the error change of transmission time, ΔU is the error change of network utilization.

As the control domain and the maximum prediction horizon are integers, the rounding of (19a) and (19b) is needed. That is the nearest integer value of the operating parameters (in actual MATLAB simulation, x is the parameter rounded: round(x)).

The role of quantization weight is quantificationally to convert the change values in parentheses of "round(x)" to the adjustment of parameters, in this section, the order of magnitude of prediction domain P, control domain M and control coefficient λ is adopted, for example, M=4, P=25, λ=0.2, the corresponding quantization weight are $\omega_1 = 1, \omega_2 = 10, \omega_3 = 0.1$.

This design, which considers factors of system control and network scheduling, will guarantee the optimization operation under the comprehensive performance of NCSs. From section 3.1, we can find that it is very important to improve the control performance of the whole system by dynamically change the network utilization in every loop and furthermore change the sampling period based on the different requirements in every loop. It adapts the system control in network environment and achieves the purpose of co-design by combined network scheduling parameters and changes the control parameters of prediction control algorithm reasonably.

4.4 General process of co-design methods

The general process of the co-design methods is (see Fig. 2):

1. Determine the plant and its parameters of NCSs.
2. Adopting GPC and EDF algorithm, defining the GPC control performance parameters and EDF scheduling parameters respectively.
3. According to the control parameters and scheduling parameters impact on system performance, design a reasonable optimization with balance between control performance and scheduling performance.
4. Use Truetime simulator to verify the system performance, then repeat the steps above if it has not meet the requirements.

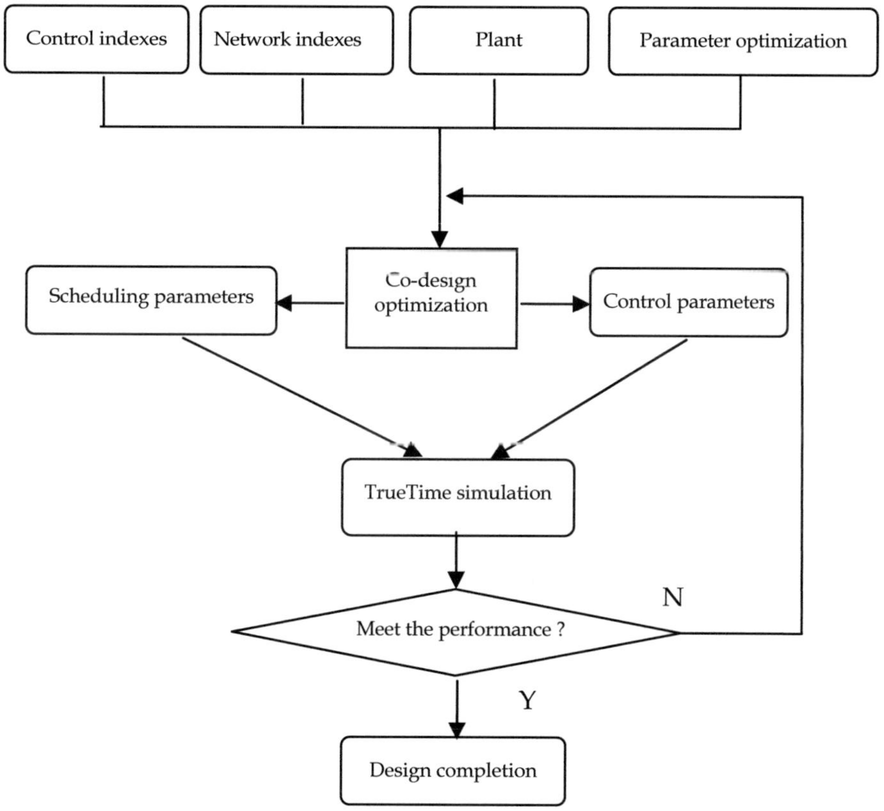

Fig. 2. General method of co-design of NCS scheduling and control

To facilitate the research of co-design, the algorithm proposed in this chapter can be extended to co-design of the other control and scheduling algorithms. And we can replace GPC with the other control algorithms and replace EDF with the other scheduling algorithms. The design idea and process are similar to the co-design algorithm presented in this chapter.

5. Simulation experiments

5.1 Simulation models and parameters' settings

In this chapter, NCS of three loops are used, the plants are the three DC (Direct Current) servo motors, and all the three loops have the same control architecture. The transfer function model of DC servo motor is:

$$G(s) = \frac{w(s)}{U_a(s)} = \frac{155.35}{s^2 + 12.46s + 11.2} \tag{20}$$

The transfer function is converted into a state-space expression:

$$\begin{cases} \dot{\boldsymbol{x}}(t) = \boldsymbol{A}\boldsymbol{x}(t) + \boldsymbol{B}\boldsymbol{u}(t) \\ \boldsymbol{y}(t) = \boldsymbol{C}\boldsymbol{x}(t) \end{cases} \tag{21}$$

$$\boldsymbol{A} = \begin{bmatrix} -12.46 & -11.2 \\ 0 & 1 \end{bmatrix}, \quad \boldsymbol{B} = \begin{bmatrix} 1 \\ 0 \end{bmatrix}, \quad \boldsymbol{C} = \begin{bmatrix} 0 & 155.35 \end{bmatrix} \circ$$

We can suppose that:

1. Sensor nodes use the time-driven, the output of the plant is periodically sampled, and sampling period is T .
2. Controller nodes and actuator nodes use event-driven.

At the sampling instant k^{th}, when the controller is event driven, after the outputs of the plant reach the controller nodes, they can be immediately calculated by the control algorithm and sent control signals, similarly, actuator nodes execute control commands at the instant of control signals arrived.

Let τ_k be the network induce delay, then

$$\tau_k = \tau_{sc} + \tau_{ca} \tag{22}$$

where τ_{sc} is the delay from sensor nodes to control nodes, τ_{ca} is the delay from control nodes to actuator nodes.

Suppose $\tau_k < T$, as the network induce delay exists in the system, the control input of the plant is piecewise constant values in a period, the control input which actuator received can be expressed by(23) (Zhang & Kong,2001):

$$\boldsymbol{v}(t) = \begin{cases} \boldsymbol{u}(k-1), & t_k < t \le t_k + \tau_k \\ \boldsymbol{u}(k), & t_k + \tau_k < t \le t_k + T \end{cases} \tag{23}$$

To discretize equation (22), and suppose the delay of NCS is stochastic, then

$$\begin{cases} \boldsymbol{x}(k+1) = \boldsymbol{A_d x}(k) + \boldsymbol{\Gamma_0 u}(k) + \boldsymbol{\Gamma_1 u}(k-1) \\ \boldsymbol{y}(k) = \boldsymbol{Cx}(k) \end{cases} \tag{24}$$

where, $\boldsymbol{A_d} = e^{AT}$, $\boldsymbol{\Gamma_0} = \int_0^{T-\tau_k} e^{As} ds\boldsymbol{B}$, $\boldsymbol{\Gamma_1} = \int_{T-\tau_k}^{T} e^{As} ds\boldsymbol{B}$.

Then introducing the augmented state vector $z(k) = [x_k^T \quad u_{k-1}^T]^T$, the above equation (24) can be rewritten as follows:

$$\begin{cases} z(k+1) = \boldsymbol{\Phi_k} z(k) + \boldsymbol{B_0 u}(k) \\ \boldsymbol{y}(k) = \boldsymbol{C_0} z(k) \end{cases} \tag{25}$$

$$\boldsymbol{\Phi_k} = \begin{bmatrix} \boldsymbol{A_d} & \boldsymbol{\Gamma_1} \\ \boldsymbol{0} & \boldsymbol{0} \end{bmatrix}, \quad \boldsymbol{B_0} = \begin{bmatrix} \boldsymbol{\Gamma_0} \\ \boldsymbol{I} \end{bmatrix}, \quad \boldsymbol{C_0} = [C \quad 0]$$

The initial sampling period $T = 10ms$, so the discretization model of DC servo motor is:

$$\begin{cases} \boldsymbol{x}(k+1) = \begin{bmatrix} 0.2625 & -0.629 \\ 0.0561 & 0.9618 \end{bmatrix} \boldsymbol{x}(k) + \begin{bmatrix} 0.0561 \\ 0.0034 \end{bmatrix} \boldsymbol{u}(k) \\ \boldsymbol{y}(k) = [0 \quad 155.35] \boldsymbol{x}(k) \end{cases} \tag{26}$$

The corresponding augmented matrix is:

$$\begin{cases} z(k+1) = \begin{bmatrix} 0.2625 & -0.629 & 0.0561 \\ 0.0561 & 0.9618 & 0.0034 \\ 0 & 0 & 0 \end{bmatrix} z(k) + \begin{bmatrix} 0 \\ 0 \\ 1 \end{bmatrix} \boldsymbol{u}(k) \\ \boldsymbol{y}(k) - [0 \quad 155.35 \quad 0\ ] z(k) \end{cases} \tag{27}$$

Convert the state space model of augment system to the CARIMA form:

$$\boldsymbol{y}(k) = 1.224\boldsymbol{y}(k-1) - 0.2878\boldsymbol{y}(k-2) + 0.5282\boldsymbol{u}(k-2) + 0.3503\boldsymbol{u}(k-3) \tag{28}$$

The simulation model structure of co-design of the networked control system with three loops is illustrated by Fig. 3. Controllers, actuators and sensors choose a Truetime kernel models respectively, the joint design optimization module in Fig.3 contains control parameter model and scheduling parameter model, and acts on the sensors and controllers of three loops, in order to optimize system operating parameters in real time.

The initial value of GPC control parameters: $M = 2$, $P = 20$, $\lambda = 0.1$, quantization weights: $\omega_1 = 1, \omega_2 = 10, \omega_3 = 0.1$; network parameters: CAN bus network, transmission rate is 800kbps, scheduling algorithm is EDF, reference input signal is step signal, amplitude is 500.

Loop1: initial sampling period $T_1 = 10ms$, size of data packet: 100bits, transmission time: $c_1 = 100 \times 8 / 800000 = 1ms$;

Loop2: initial sampling period $T_2 = 10ms$, size of data packet: 90bits, transmission time: $c_2 = 0.9ms$;

Loop3: initial sampling period $T_3 = 10ms$, size of data packet: 80bits, transmission time: $c_3 = 0.8ms$.

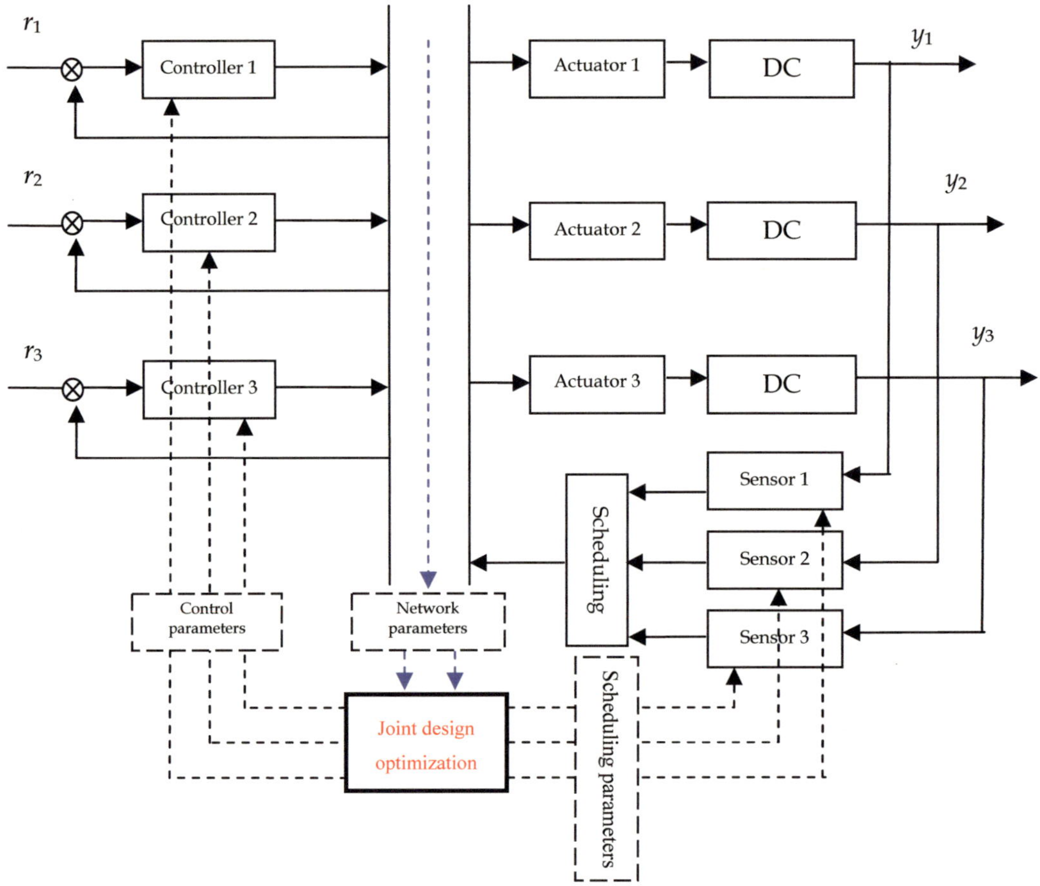

Fig. 3. Simulation framework of NCS with three loops

5.2 Simulation experimental results and their analyses

The following is comparison of joint design and no joint design, in order to facilitate comparison and analysis, defining as follows: "Co-design" expresses the simulation curve of joint design, while "N-Co-design" expresses the no joint design. Network induce delay can be achieved by delay parameter "exectime" in Truetime simulation. Node 1, 2 and 3 indicate the actuator, controller and sensor in loop 1 respectively; Node 4, 5 and 6 indicate the actuator, controller and sensor in loop 2 respectively; Node 7, 8 and 9 indicate the actuator, controller and sensor in loop 3 respectively.

Case 1: In the absence of interfering signals, and network induce delay is $\tau_k = 0ms$, under ideal conditions, the system response curves of both algorithms are shown in fig.4, where number 1, 2, 3 denote the three loops respectively.

From Fig. 4, in the situation of without interference and delay, the system response curves of Co-design and N-Co-design system response curves are basically consistency; they all show

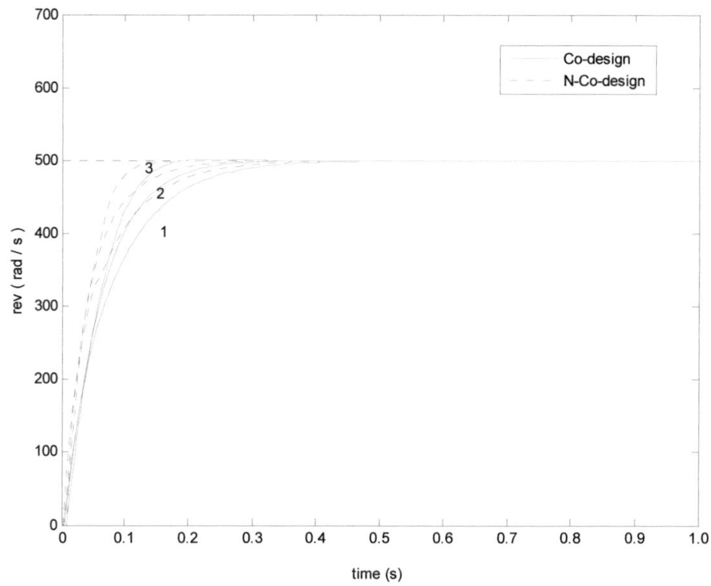

Fig. 4. The system response

the better performance. The system performance of N-Co-design is better than the Co-design one in terms of the small rise time and faster dynamic response. The main reason is the large amount of computation of GPC, and the system adds the amount of computation after considering Co-design, these all increase the complexity of the system and computation delay of network. So, in the ideal case, the N-Co-design system has the better performance.

Case 2: Interference signal network utility is 20%, and network induce delay is $\tau_k = 3ms$, τ_k is bounded by 0 and 1/2 of sampling period, that is 0~5ms. At this case, the network environment is relatively stable, network-induce delay is relatively small, interference signal occupied relatively small bandwidth.

Network scheduling timing diagrams of the two algorithms are shown as Fig. 5 and Fig. 6.

From the scheduling time diagrams of Co-design and N-Co-design (Fig.5 and Fig. 6), we can find that data transmission condition are better under two algorithms for loop1 and loop2, there are no data conflict and nonscheduled situation. But for loop3, compared with the co-design system, the N-Co-design shows the worse scheduling performance and more latency situations for data transmission and longer duration (longer than 7ms, sometimes), this greatly decreases the real-time of data transmission. The Co-design system shows the better performance: good real-time of data transmission, no latency situations for data, which corresponds to shorter adjustment time for loop3 in Fig.7. The system response curves are shown in Fig. 7.

Fig.7 shows that when the changes of network induce delay are relatively small, the response curves of co-design system and N-Co-design system are basically consistency, all three loops can guarantee the system performance. The system performance of N-Co-design is better than the Co-design one in terms of the small rise time and faster dynamic response.

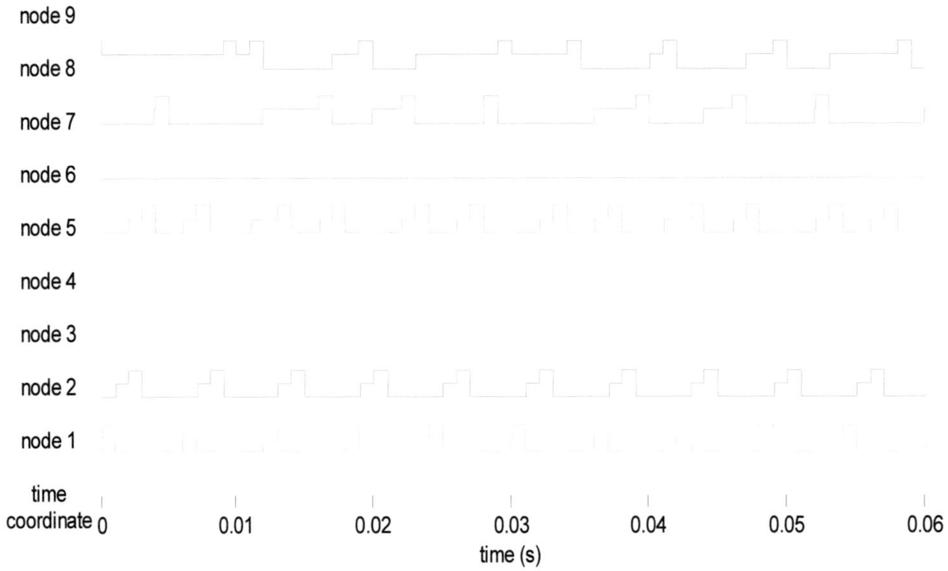

Fig. 5. The network scheduling time order chart of N-Co-design

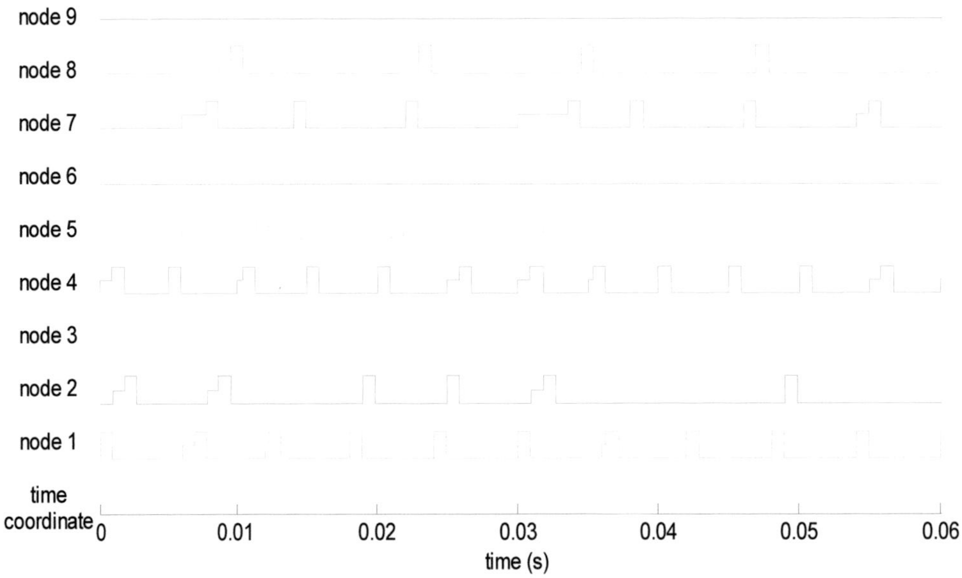

Fig. 6. The network scheduling time order chart of Co-design

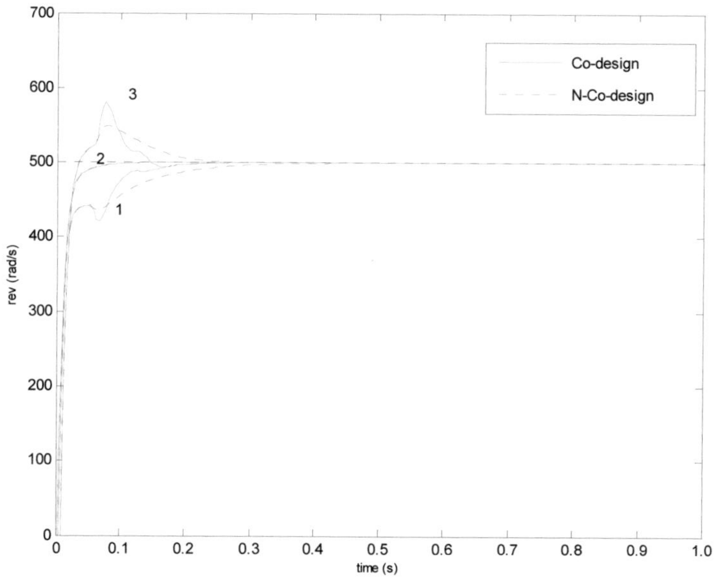

Fig. 7. The system response

The main reason is the large amount of computation of GPC, and the system adds the amount of computation after considering Co-design, these all increase the complexity of the system and computation delay of network. So, in smaller delay or less network load situations, the N-Co-design system has the better performance.

Case 3: Interference signal network utility is 40%, and network induce delay is $\tau_k = 8ms$, τ_k is smaller than the sampling period 10ms. At this case, the network environment is relatively worse, interference signal occupied relatively big bandwidth, network-induce delay is relatively big.

Network scheduling timing diagrams of the two algorithms are shown as Fig. 8 and Fig. 9.

From the two situations (Figure 8 and Figure 9) we can see that the data transmission condition of Co-design system is better than the N-Co-design one with all the three loops. Although there are no data conflictions and nonscheduled situation, the N-Co-design system shows the worse scheduling performance and more situations of latency data, which greatly affect the real-time data. This is bad for the real-time networked control system. In contrast, the Co-design system is better, latency data is the less, which can achieve the performance of effectiveness and real-time for the data transmission.

As shown in system response curves (Fig. 10) and scheduling timing diagrams (Fig. 8 and Fig. 9), when the network induce delay is bigger, the three loops of Co-design denote the better control and scheduling performance: better dynamic response, smaller overshoot, less fluctuation; scheduling performance guarantees the network induce delay no more than the sampling period, data transfer in an orderly manner, no nonscheduled situation. So, under the case of worse network environment and bigger network induce delay, the system with co-design expresses the better performance, while the worse performance of the system of N-Co-design. The main reason is the operation of control algorithm of Co-design with

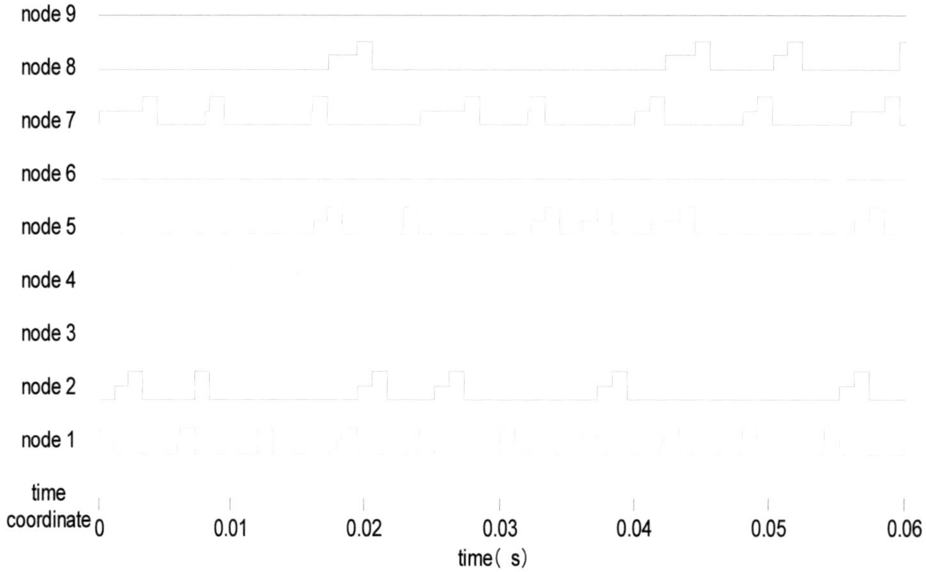

Fig. 8. The network scheduling time order chart of N-Co-design

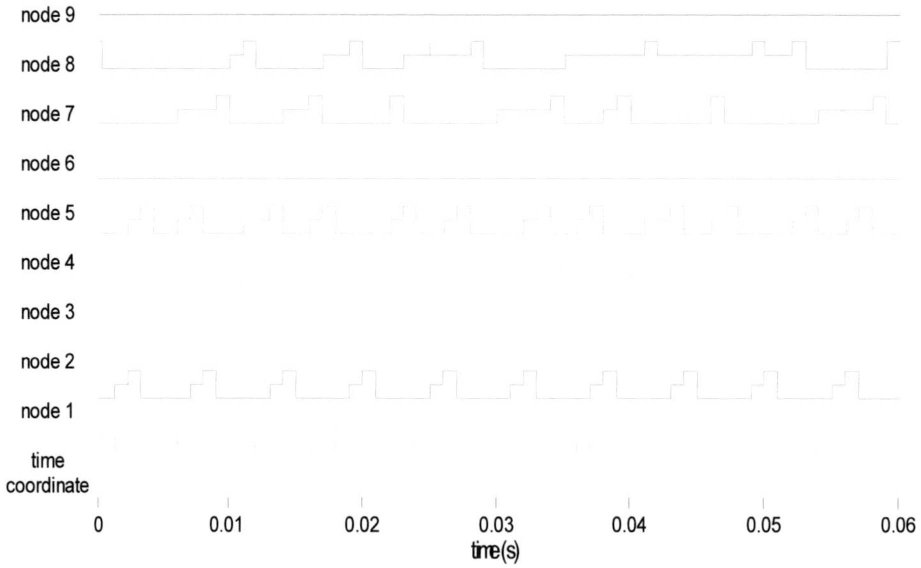

Fig. 9. The network scheduling time order chart of Co-design

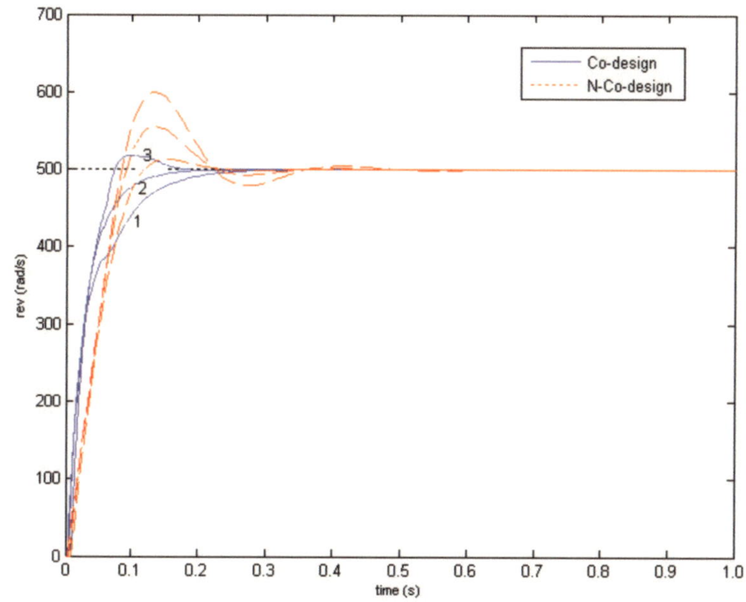

Fig. 10. The system response

considering the effect of network. When the network impact increases, the effect is decreased on the control algorithm.

Case 4: To illustrate the superiority and robustness of the designed algorithm, we add interference to the system at the instant t=0.5s, that is increasing the network load suddenly, the network utility of interference increases from 0 to 40%. The system response curves of the three loops with the two algorithms are shown as follows.

From the system response curves, we can see that the system of Co-design shows the better robustness and faster dynamic performance when increasing interference signal suddenly. In loop 1 (Fig. 11), the system pulse amplitude of Co-design is small, the rotational speed amplitude is 580rad/s (about 5400 cycles/min), the rotational speed amplitude of N-Co-design is nearly 620 rad/s; in loop 2 (Fig. 12), the system amplitude and dynamic response time increase compared to loop 1, but the both can guarantee the normal operation of system; but in loop 3 (Fig. 13), the system occurs bigger amplitude (nearly 660 rad/s) and longer fluctuation of N-Co-design system after adding interference signal, and also the slower dynamic response. The system of Co-design shows the better performance and guarantees the stable operation of system.

From the four cases above, we can conclude that under the condition of better network environment, the system performance of Co-design is worse than the one without Co-design, this is because the former adopts GPC algorithm, and GPC occupies the bigger calculation time, it further increases the complexity of the algorithm with joint design optimization. So, under the ideal and small delay condition, the system without Co-design is better, contrarily, the Co-design is better. When adding interference signal suddenly, the system with Co-design shows the better network anti-jamming capability and robustness.

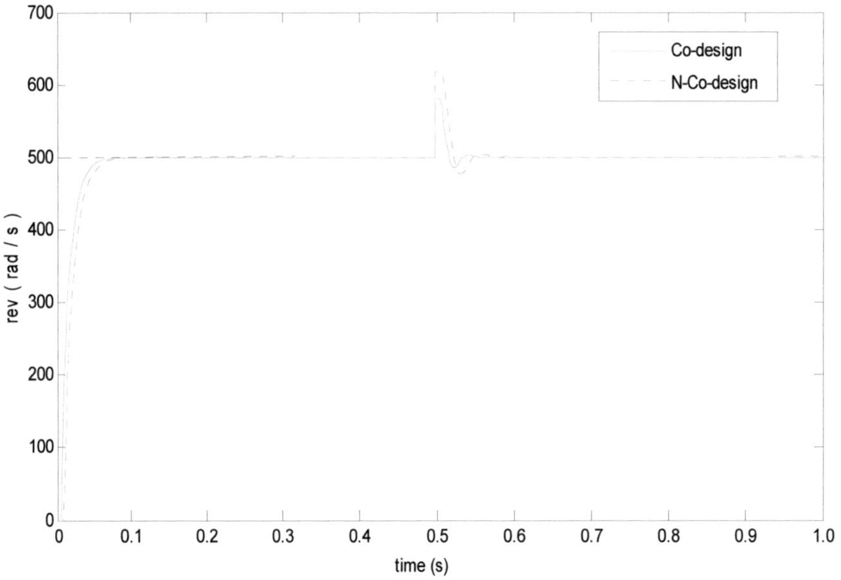

Fig. 11. The system response of Loop 1

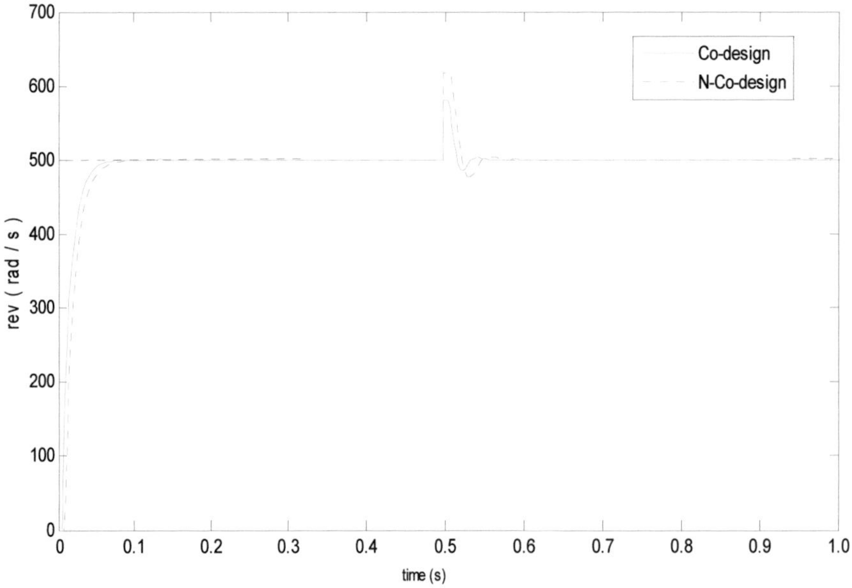

Fig. 12. The system response of Loop 2

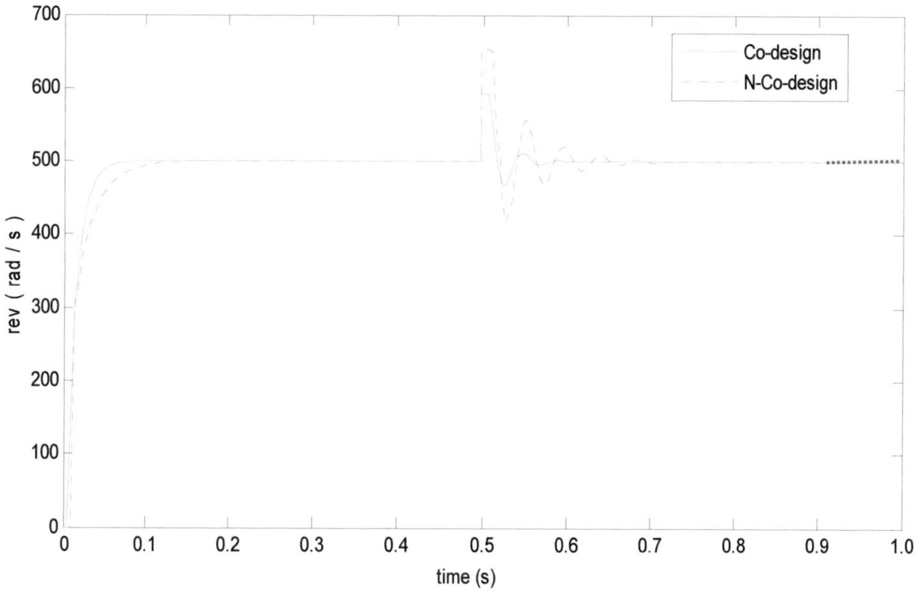

Fig. 13. The system response of Loop 3

6. Conclusion

First introducing the theory and parameters of GPC , then the EDF scheduling algorithm and parameter are presented. The co-design of control and scheduling is proposed after analyzing the relationship between predictive control parameters and scheduling parameters for a three-loop DC servo motor control system. By analyzing the effect on system performance by the control parameters and the scheduling parameters, a joint optimization method is designed considering the balance between control performance and scheduling performance. Finally this algorithm is validated by Truetime simulation, in the cases of big delay and bad environment, especially the presence of external interference, the co-design system shows the better performance, such as good robustness and anti-jamming capability.

7. Acknowledgment

This work is supported in part by National Natural Science Foundation of China (NSFC) under Grant No.60872012

8. References

Gaid M B,Cela A,Hamam Y. (2006). Optimal integrated control and scheduling of systems with communication constraints, *Proceedings of the Joint 44th IEEE Conference on Decision and Control and European Control Conference*, pp. 854-859, ISBN 0-7803-9567-0, Seville, Spain. December, 2005

Gaid M B, Cela A, Hamam Y. (2006). Optimal integrated control and scheduling of networked control systems with communication constraints: application to a car suspension system. *IEEE Transactions on Control Systems Technology,* Vol.14, No.4, (July 2006), pp.776-787, ISSN 1063-6536

Arzen K E, Cervin A, Eker J. (2000). An introduction to control and scheduling co-design. *Proceedings of the 39th IEEE Conference on Decision and Control,*pp. 4865-4870,ISSN 0191-2216, Sydney, Australia. Decmber, 2000

Calrke, Mohtadi. (1989). Properties of Generalized Predictive Control, *Automatic,* Vol.25, No.6,(November 1989), pp.859-875, ISSN 0005-1098

Wei Wang et al.(1998). *Generalized predictive control theory and its application,* Science Press, ISBN 978-7-030-06804-0, Beijing

Shouping Guan, Wei Zhou.(2008). *Networked control system and its application,* Publishing house of electronics industry, ISBN 978-7-121-06946-8, Beijing

Baocang Ding. (2008). *Predictive control theory and methods,* China machine press, ISBN 978-7-111-22898-1, Beijing

Bin Li.(2009). Study of fuzzy dynamic scheduling and variable sampling period algorithm of NCS. *Beijing Jiaotong University,*Vol.33, No.2, (April 2009),pp.98-102, ISSN 1673-0291

Lian Feng-li, Moyne James, Tilbury Dawn. (2001). Time Delay Modeling and Sampling Time Selection for Networked Control Systems, *Proceedings of the ASME Dynamic Systems and Control,* pp. 1265-1272, New York, USA. January, 2001

Xuelin Zhang, Feng Kong.(2008). Research on Scheduling of Networked Control Systems Based on CAN Bus, *Control and automation publication group,*ISSN 1008-0570

Pedreiras P, Almeida L. (2002). EDF Message Scheduling on Controller Area Network, *Computing & Control Engineering Journal,* Vol. 13, No. 4, (August 2002), pp. 163-170

Mayne, Rawlings, Ral and Scokaert. (2003). Constrained Model Predictive Control Stability and Optimality, *Automatic,*Vol.36, No.6, pp.789-814, ISSN 0005-1098

Yaohua Hu, Suwu Xu, Gongfu Xie et al. (2000). Robustness of General Prediction Control, *Journal of Dalian Maritime University,* Vol. 26, No. 1, (Feburary 2000), pp. 75-79, ISSN 1006-7736

Zengqiang, Chen, Xinhua Wang, Zhuzhi Yuan. (2003). GPC Control and Analysis of Stability for Hybrid Systems,*Computing Technology and Automation,* Vol. 22, No. 2,(June 2003), pp. 1-4, ISSN 1003-6199

5

Predictive Control for the Grape Juice Concentration Process

Graciela Suarez Segali[1] and Nelson Aros Oñate[2]
[1]*Department of Chemical Engineering, Faculty of Engineering, National University of San Juan, Avda. Libertador San Martín, San Juan,*
[2]*Department of Electrical Engineering, Faculty of Engineering, University of La Frontera, Avda. Francisco Salazar, Temuco,*
[1]*Argentina*
[2]*Chile*

1. Introduction

Concentrated clear grape juices are extensively used in the enological industry. Their use as constituents of juices, jellies, marmalades, jams, colas, beverages, etc., generates a consumer market with an increasing demand because they are natural products with an industrial versatility that allows them to compete with other fruit juices.

Argentina is one of the principal producers and exporters of concentrated clear grape juices in the world. They are produced mainly in the provinces of Mendoza and San Juan (Argentine Republic) from the virgin grape juice and in the most part from sulfited grape juices. The province of Mendoza's legislation establishes that a portion of the grapes must be used for making concentrated clear grape juices. This product has reached a high level of penetration in the export market and constitutes an important and growing productive alternative.

An adequate manufacturing process, a correct design of the concentrate plants and an appropriate evaluation of their performance will facilitate optimization of the concentrated juices quality parameters (Pilati, 1998; Rubio, 1998). The plant efficiency is obtained from knowledge of the physics properties of the raw material and products (Moressi, 1984; Piva, 2008). These properties are fundamental parameters that are used in the designing and calculations on all the equipment used and also in the control process.

The juices (concentrate and intermediate products) physical properties, such as density, viscosity, boiling point elevation, specific heat and coefficient of thermal expansion, are affected by their solid content and their temperature (Schwartz, 1986). For this reason, it is necessary to know the physical properties values, as a function of the temperature and the solids content, during the manufacture process, not just to obtain an excellent quality, but also to develop a data base, that is essential for optimizing the installation design and the transformation process itself. The principal solids constituents of clear grape juices are sugars (mostly glucose and fructose) and its concentration affects directly the density, viscosity and refraction index.

The type and magnitude of degradation products will depend on the starting reagent condition (Gogus, et al., 1998). Acetic, formic, and D/L-lactic acids were identified at the end of thermal degradation of sugar solutions (Asghari and Yoshida, 2006), and a reaction scheme was proposed by Ginz et al. (2000). Sugar degradation may result in browning of solutions with polymeric compounds as the ultimate product of degradation, generally known as "melanoidins", involving the formation of 5-(hydroxymethyl)-2-furancarboxaldehyde (5-HMF) as intermediate.

Barbieri and Rossi (1980) worked with white concentrated clear grape juice in a falling film multiple effect evaporators. They obtained 18.2, 27.3, 38.6, 48.6 and 64.6 °Brix samples. They measured density, viscosity and boiling point elevation as a function of soluble solids concentration and temperature. They presented the results in plots with predictive equations for the properties studied.

Di Leo (1988) published density, refraction index and viscosity data for a rectified concentrated grape juice and an aqueous solution of a 1:1 glucose/levulose mixture, for a soluble solids concentrate range from 60 to 71% (in increments of 0.1%) and 20 °C. The author determinated the density in undiluted and 2.5-fold diluted samples (100 g of clear grape juice in 250 ml of solution at 20 °C), finding different results between both determinations. He recommended measuring density without dilution.

Pandolfi et al., (1991) studied physical and chemical characteristics of grape juices produced in Mendoza and San Juan provinces, Argentina. They determined density at 20°C in sulfited grape juices of 20–22°Bx and concentrated grape juices of 68–72°Bx. They obtained no information on intermediate concentrations or other temperatures. In general, the clarified juice concentrates have a Newtonian behavior (Ibarz & Ortiz, 1993; Rao, Cooley & Vitali, 1984; Sáenz & Costell, 1986; Saravacos, 1970).

Numerous industrial processes are multivariable systems which require a large number of variables to be controlled simultaneously (Kam, 1999; Kam, 2000). The controller design is for this type of system has a great interest in control theory (Doyle, 1979; Freudenberg, 1988; Friedland, 1989; Middleton, 1990; Zang, 1990; Aros, 2008; Suarez, 2010). This work presents an interactive tool to facilitate understanding of the control of multivariable systems (MIMO) using the technique of Generalized Predictive Control (GPC). The tool can handle the main concepts of predictive control with constraints and work both as monovariable and multivariable systems.

The GPC for systems multivariable, MBPC or Model Based Predictive Control includes a set of techniques to cover wide range of problems from those with relatively simple dynamics to other more complexes (unstable, large delays, nonminimum phase systems, etc.). Among its many advantages (Camacho & Bordons, 1999) is its easy adaptation to multivariable systems. One of the most important techniques in academia for predictive control is the Generalized Predictive Control (Clarke et al., 1987). The characteristic of this strategy, as shown in figure 1, is that at each sampling time and using a process model, predicting the future outputs for a given horizon. With these predicted outputs, using an objective function and taking into account the constraints that affect the process (eg on the inputs and outputs) are calculated future control increments. Finally, we apply the first control signal is calculated, the rest is discarded and the horizon moves forward, repeating the calculations in the next sampling period (receding horizon strategy).

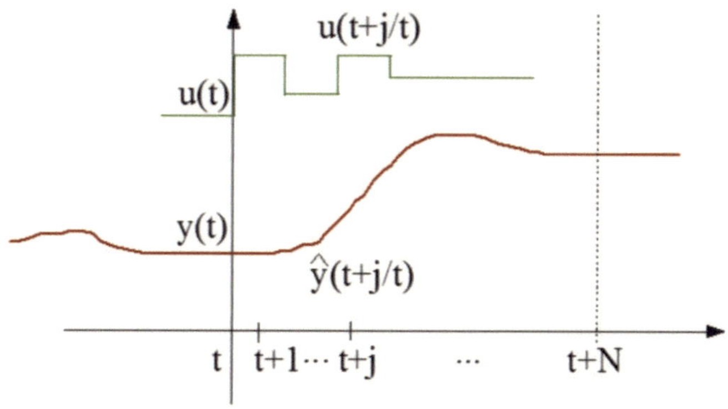

Fig. 1. MBPC action.

The GPC technique is based on the use of models derived from transfer functions (transfer matrices in the multivariate case). The use of a formulation of this kind against an internal description has certain advantages in the field of development of interactive tools. The transfer function formulation is more intuitive, being based only on information input and output measurable and arrange its elements (poles and zeros) of a clear physical meaning and interpretation.

This is critical in the design of interactive tools, which simultaneously shows different representations of the system that allow to analyze how the change affects any parameter of the plant-controller-model global behavior of the controlled system without ever losing its physical sense, allowing to develop their intuition and skills.

The basic idea was proposed of GPC is to calculate a sequence of future control signals in such a way that it minimizes a multistage cost function defined over a prediction horizon. The index to be optimized is the expectation of a quadratic function measuring the distance between the predicted systems output and some predicted reference sequence over the horizon plus a quadratic function measuring the control effort. This approach was used in Lelic & Wellstead (1987) and Lelic & Zarrop (1987), to obtain a generalized pole placement controller which is an extension of the well-known pole placement controllers Allidina & Hughes (1980) and belongs to the class of extended horizon controllers.

Generalized Predictive Control has many ideas in common with the predictive controllers previously mentioned since it is based upon the same concepts but it has some differences. As will be seen, it provides an analytical solution (in the absence of constraints)nit can deal with unstable and nonminimum phase plants and it incorporates the concept of control horizon as well as the consideration of weighting control increments in the cost function. The general set of choices available for GPC leads to a greater variety of control objectives compared to other approaches, some of which can be considered as subsets or limiting cases of GPC. In particular, the strategy GPC uses the model CARIMA (Controlled Auto Regressive Integrated Moving Average) to predict the process output.

2. Process description

Figure 2 show the input and output streams in a vertical generic effect evaporator with long tubes. The solution to be concentrated circulates inside the tubes, while the steam, used to heat the solution, circulates inside the shell around the tubes.

The evaporator operates in co-current. The solution to be concentrated and the steam are fed to the first effect by the bottom and by the upper section of the shell, respectively. Later on, the concentrated solution from the first effect is pumped to the bottom of the second effect, and so on until the fourth effect. On the other hand, the vapor from each effect serves as heater in the next one. Finally, the solution leaving the fourth effect attains the desired concentration.

Each effect has a baffle in the upper section that serves as a drops splitter for the solution dragged by the vapor. The vapor from the fourth effect is sent to a condenser and leaves the process as a liquid. The concentrated solution coming from the fourth effect is sent to a storage tank.

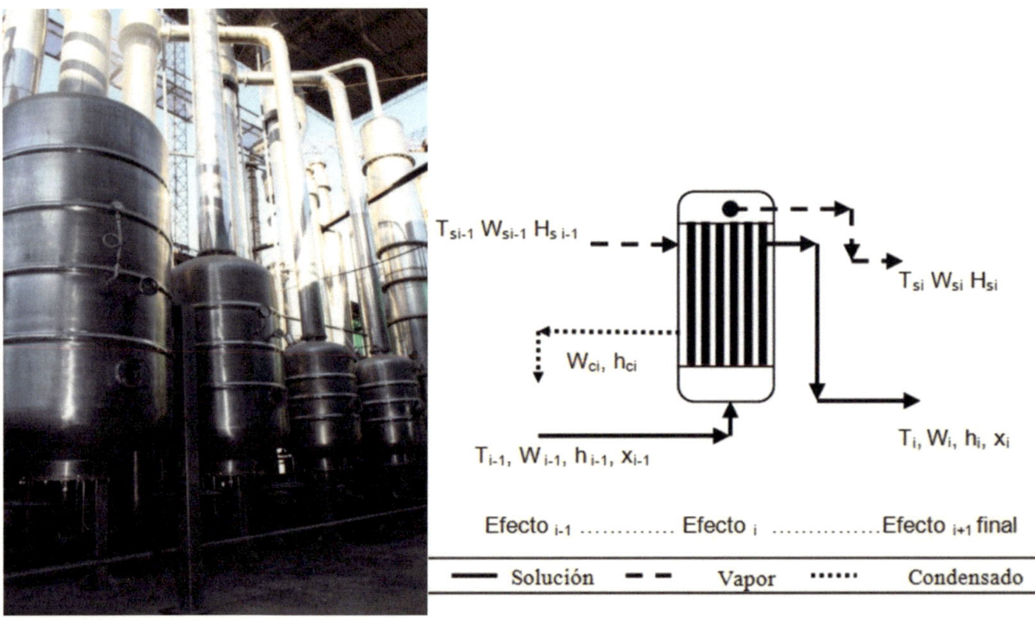

Fig. 2. Photo of evaporator and scheme of effect i in the four-stage evaporator flow sheet. $i = 1, \cdots, 4$.

3. Phenomenological model

Stefanov & Hoo (2003) have developed a rigorous model with distributed parameters based on partial differential equations for a falling-film evaporator, in which the open-loop stability of the model to disturbances is verified. On the other hand, various methods have been proposed in order to obtain reduced-order models to solve such problems (Christofides, 1998; El-Farra, Armaou and Christofides, 2002; Hoo and Zheng, 2001; Zheng and Hoo, 2002). However, the models are not a general framework yet, which assure an effective implementation of a control strategy in a multiple effect evaporator.

In practice, due to a lack of measurements to characterize the distributed nature of the process and actuators to implement such a solution, the control of systems represented by partial differential equation (PDE) in the grape juice evaporator, is carried out neglecting the spatial variation of parameters and applying lumped systems methods. However, a distributed parameters model must be developed in order to be used as a real plant to test advance control strategies by simulation.

In this work, it is used the mathematical model of the evaporator developed by Ortiz *et al.* (2006), which is constituted by mass and energy balances in each effect. The assumptions are: the main variables in the gas phase have a very fast dynamical behavior, therefore the corresponding energy and mass balances are not considered. Heat losses to surroundings are neglected and the flow regime inside each effect is considered as completely mixed.

a. Global mass balances in each effect:

$$\frac{dW_i}{dt} = W_{i-1} - W_{si} - W_i \tag{1}$$

in this equations $W_i, i = 1,...,4$ are the solution mass flow rates leaving the effects 1 to 4, respectively. W_0 is the input mass flow rate that is fed to the equipment. $W_{si}, i = 1,...,4$ are the vapor mass flow rates coming from effects 1 to 4, respectively. $dMi / dt, i = 1,...,4$ represent the solution mass variation with the time for each effect.

b. Solute mass balances for each effect:

$$\frac{d(W_i X_i)}{dt} = W_{i-1} X_{i-1} - W_i X_i \tag{2}$$

where, $X_i, i = 1,...,4$ are the concentrations of the solutions that leave the effects 1 to 4, respectively. X_o, is the concentration of the fed solution.

c. Energy balances:

$$\frac{dW_i h_i}{dt} = W_{i-1} h_{i-1} - W_i h_i - W_{si} H_{si} + A_i U_i (T_{si-1} - T_i) \tag{3}$$

where, $h_i, i = 1,...,4$ are the liquid stream enthalpies that leave the corresponding effects, h_0 is the feed solution enthalpy, and $H_{si}, i = 1,...,4$ are the vapor stream enthalpies that leave the corresponding effects and, A_i represents the heat transfer area in each effect. The model also includes algeb raic equations. The vapor flow rates for each effect are calculated neglecting the following terms: energy accumulation and the heat conduction across the tubes. Therefore:

$$W_{si} = \frac{U_i A_i (T_{si-1} - T_i)}{H_{si-1} - h_{ci}} \tag{4}$$

For each effect, the enthalpy can be estimated as a function of temperatures and concentrations (Perry, 1997). Them:

$$H_{si} = 2509.2888 + 1.6747 T_{si} \tag{5}$$

$$h_{ci} = 4.1868 T_{si} \tag{6}$$

$$C_{pi} = 0.80839 - 4.3416 \cdot 10^{-3} X_i + 5.6063 \cdot 10^{-4} T_i \quad (7)$$

$$h_i = 0.80839 T_i - 4.316 \cdot 10^{-3} X_i T_i + 2.80315 \cdot 10^{-4} T_i^2 \quad (8)$$

$T_i, i = 1,\ldots,4$ are the solution temperatures in each effect, and T_{s0}, is the vapor temperature that enters to the first effect. $T_{si}, i = 1,\ldots,4$ are the vapor temperatures that leave each effect.

The heat transfer coefficients are:

$$U_i = \frac{490.D^{0.57} W_{si}^{3.6JL}}{\mu_i^{0.25} \Delta T_i^{0.1}} \quad (9)$$

Once viscosity values were established at different temperatures, (apparent) flow Activation Energy values for each studied concentration were calculated using the Arrhenius equation:

$$\mu = \mu_\infty \exp(-\frac{E_a}{RT}) \quad (10)$$

$$\mu_\infty = -\exp(a_0 + a_1 Brix + a_2 Brix^2 \quad (11)$$

$$E_a/_T = -\exp(a_0 + a_1 Brix + a_2 Brix^2 \quad (12)$$

The global heat-transfer coefficients are directly influenced by the viscosity and indirectly by the temperature and concentration in each effect. The constants a_0, a_1 y a_2 depend on the type of product to be concentrated (Kaya, 2002; Perry, 1997; Zuritz, 2005).

Although the model could be improved, the accuracy achieved is enough to incorporate a control structure.

4. Standard model predictive control

The biggest problem that arises in the implementation of conventional PID controllers, arises when there are high nonlinearities and long delays, a possible solution to these arises with the implementation of predictive controllers, in which the entry in a given time (t) will generate an output at a time (t +1), using a control action at time t.

The model-based predictive control is currently presented as an attractive management tool for incorporating operational criteria through the use of an objective function and constraints for the calculation of control actions. Furthermore, these control strategies have reached a significant level of acceptability in practical applications of industrial process control.

The model-based predictive control is mainly based on the following elements:

- The use of a mathematical model of the process used to predict the future evolution of the controlled variables over a prediction horizon.
- The imposition of a structure in the future manipulated variables.
- The establishment of a future desired trajectory, or reference to the controlled variables.
- The calculations of the manipulated variables optimizing a certain objective function or cost function.
- The application of control following a policy of moving horizon.

4.1 Generalized predictive control

The CARIMA model of the process is given by:

$$A(z^{-1})y(t) = B(z^{-1})u(t-1) + \frac{1}{\Delta}C(z^{-1})e(t) \tag{13}$$

with

$$\Delta = 1 - z^{-1}$$

And the C polynomial is chosen to be 1, from what they if C^{-1} can be truncated it can be absorbed into A and B.

The GPC algorithm consists of applying a sequence that minimizes a multistage cost function of the form

$$J(N_1, N_2, N_u) = \sum_{j=N_1}^{N_2} \delta(j)[\hat{y}(t+j|t) - w(t+j)]^2 + \sum_{j=1}^{N_1} \lambda(j)[\Delta u(t+j-1)]^2 \tag{14}$$

where:

$\hat{y}(t+j|t)$ is a sequence of (j) best predictions from the output of the system later instantly t and performed with the known data to instantly t.

$\Delta u(t+j-1)$ is a sequence control signal increases to come, to be obtained from the minimization of the cost function.

N_1, N_2 and N_u are the minimum and maximum costing horizons, and control horizon. N_1 and N_2 That does not necessarily coincide with the maximum prediction horizon. The meaning of them is quite intuitive, they mark the limits of the moments that criminalizes the discrepancy of the output with the reference.

δ(j) and λ(j) are weighting factors they are sequences are respectively weighted tracking errors and future control efforts. Usually considered constant values or exponential sequences. These values can be used as tuning parameters.

Reference trajectory: one of the benefits of predictive control is that if you know a priori the future evolution of the reference, the system can start to react before the change is actually carried out, avoiding the effects of the delay in the response of the process. On the criterion of minimizing (Bitmead et al., 1990), most of the methods often used a trajectory of reference w(t+j) which does not necessarily coincide with the actual reference. Normally it would be a soft approach from the current value of the output y (t) to the known reference, through a first-order dynamics.

$$w(t+j) = \alpha w(t+k-1) + (1-\alpha)r(t+j) \tag{15}$$

where

α is a parameter between 0 and 1 that constitutes an adjustable value that will influence the dynamic response of the system. where $\alpha = \text{diag}(\alpha_1, \alpha_2, \ldots, \alpha_n)$ is the diagonal soften factor matrix;

$(1-\alpha) = \text{diag}(1-\alpha_1, 1-\alpha_2, \ldots 1-\alpha_n)$; r(t+j) is the system's future set point sequence. By employing this cost function, the distance between the model predictive output and the soften future set point sequence is minimized over the predictive horizon while the variation of the control input is preserved small over the control horizon.

In order to optimize the cost function the optimal prediction of y(t+j) for $j \geq N_1$ and $j \leq N_2$ will be obtained. Consider the following Diophane equation:

$$1 = E_j(z^{-1})\tilde{A}(z^{-1}) + z^{-1}F_j(z^{-1}) \quad (16)$$

where $\tilde{A}(z^{-1}) = \Delta A(z^{-1})$

The polynomial E_j and F_j are uniquely defined with degrees j-1 and na, respectively. They can be obtained by dividing 1 by $\tilde{A}(z^{-1})$ until the remainder can be factorized as $z^{-1} F_j(z^{-1})$. The quotient of the division is the polynomial E_j (z^{-1}).

$$\tilde{A}(z^{-1})E_j(z^{-1})y(t+j) = E_j(z^{-1})B(z^{-1})\Delta u(t+j-d-1) + E_j(z^{-1})e(t+j) \quad (17)$$

Considering the equation (16), the equation (17) can be written as

$$\left(1 - z^{-1}F_j(z^{-1})\right)y(t+j) = E_j(z^{-1})B(z^{-1})\Delta u(t+j-d-1) + E_j(z^{-1})e(t+j)$$

which can be rewritten as:

$$y(t+j) = F_j(z^{-1})y(t) + E_j(z^{-1})B(z^{-1})\Delta u(t+j-d-1) + E_j(z^{-1})e(t+j) \quad (18)$$

As the degree of polynomial E_j (z^{-1}) = j-1the noise terms in equation (18) are all in the future. The best prediction of y (t+j) is therefore:

$$\hat{y}(t+|t) = G_j(z^{-1})\Delta u(t+j-d-1) + F_j(z^{-1})y(t) \quad (19)$$

Where $G_j(z^{-1}) = E_j(z^{-1})B(z^{-1})$

There are other ways to formulate a GPC as can be seen in Albertos & Ortega, (1989)

The polynomials E_j, F_j and G_j can be obtained recursively.

$$F_j(z^{-1}) = F_{j,0} + F_{j,1}(z^{-1}) + F_{j,2}(z^{-2}) + \cdots + F_{j,na}(z^{-na})$$

$$E_j(z^{-1}) = E_{j,0} + E_{j,1}(z^{-1}) + E_{j,2}(z^{-2}) + \cdots + E_{j,j-1}\left(z^{-(j-1)}\right)$$

$$G_j(z^{-1}) = G_{j,0} + G_{j,1}(z^{-1}) + G_{j,2}(z^{-2}) + \cdots + G_{j,j-1}\left(z^{-(j-1)}\right)$$

for instant j +1

$$F_{j+1}(z^{-1}) = F_{j+1,0} + F_{j+1,1}(z^{-1}) + F_{j+1,2}(z^{-2}) + \cdots + F_{j+1,na}(z^{-na})$$

$$E_{j+1}(z^{-1}) = E_j(z^{-1}) + E_{j+1,j}\left(z^{-j}\right)$$

$$G_{j+1}(z^{-1}) = G_j(z^{-1}) + F_{j,0}\left(z^{-j}\right)B$$

Consider the group of j ahead optimal prediction For a reasonable response, these bounds are assumed to be Camacho & Bordons, (2004):

$$N_1 = d + 1$$

$$N_2 = d + N$$

$$N_u = N$$

$$y = Gu + F(z^{-1}) + G'(z^{-1})\Delta u(t-1) \tag{20}$$

$$y = \begin{bmatrix} \hat{y}(t+d+1|t) \\ \hat{y}(t+d+2|t) \\ . \\ . \\ \hat{y}(t+d+N|t) \end{bmatrix} \quad G = \begin{bmatrix} G_0 & 0 \ldots & 0 \\ G_1 & G_0 \ldots & 0 \\ & . & \\ & . & \\ G_{N-1} & G_{N-2} & G_0 \end{bmatrix} \quad u = \begin{bmatrix} \Delta u(t) \\ \Delta u(t+1) \\ . \\ . \\ \Delta u(t+N-1) \end{bmatrix}$$

$$F(z^{-1}) = \begin{bmatrix} F_{d+1}(z^{-1}) \\ F_{d+2}(z^{-1}) \\ . \\ . \\ F_{d+N}(z^{-1}) \end{bmatrix} \quad G'(z^{-1}) = \begin{bmatrix} (G_{d+1}(z^{-1}) - G_0)z \\ (G_{d+2}(z^{-1}) - G_0 - G_1 z^{-1})z^2 \\ . \\ . \\ (G_{d+N}(z^{-1}) - G_0 - G_1 z^{-1} - \cdots G_{N-1} z^{-(N-1)})z^N \end{bmatrix}$$

After making some assumptions and mathematical operations the equation (14) is written:

$$J = (Gu + f - w)^T (Gu + f - w) + \lambda u^T u \tag{21}$$

where

$$f = G'(z^{-1})\Delta u(t-1)$$

$$w = [w(t+d+1) w(t+d+2) \ldots w(t+d+N)]^T$$

Then (21) is

$$J = \frac{1}{2} u^T H u + b^T u + f_0$$

with

$$H = 2(G^T G + \lambda I)$$

$$b^T = 2(f - w)^T G$$

$$f_0 = (f - w)^T (f - w)$$

Many processes are affected by external disturbances caused by variation of variables that can be measured. Consider, for example, the evaporated where the first effect temperature is controlled by manipulating the steam of temperature, any variation of the steam temperature, influence the first effect temperature. These type of perturbations, also known as load disturbances, can easily be handled by the use of feedforward controllers. Known disturbances can be taken explicitly into account in MBPC, as will be seen in the following.

Consider a process described by the following in this case the CARIMA model must be changed to include the disturbances:

$$A(z^{-1})y(t) = B(z^{-1})u(t-1) + D(z^{-1})v(t) + \frac{1}{\Delta} C(z^{-1})e(t) \tag{22}$$

Where the variable v(t) is the measured disturbance at time t and $D(z^{-1})$ is a polynomial defined as:

$$D(z^{-1}) = d_0 + d_1 z^{-1} + d_2 z^{-2} + \cdots + d_{nd} z^{-nd}$$

If equation (16) is multiplied by $\Delta E_j(z^{-1}) z^j$.

$$E_j(z^{-1})\tilde{A}(z^{-1})y(t+j) = E_j(z^{-1})B(z^{-1})\Delta u(t+j-1)$$
$$+E_j(z^{-1})D(z^{-1})\Delta v(t+j) + E_j(z^{-1})e(t+j)$$

and manipulation these equation, we get

$$y(t+j) = F_j(z^{-1})y(t) + E_j(z^{-1})B(z^{-1})\Delta u(t+j-1)$$
$$+E_j(z^{-1})D(z^{-1})\Delta(t+j) + E_j(z^{-1})e(t+j)$$

Notice that because the degree of $E_j(z^{-1})$ is j-1, the noise terms are all in the future; by taking the expectation operator and considering that E[e(t)] = 0 the expected value for y (t+j) is given by:

$$\hat{y}(t+j|t) = E[y(t+j)]$$
$$= F_j(z^{-1})y(t) + E_j(z^{-1})B(z^{-1})\Delta u(t+j-1) + E_j(z^{-1})D(z^{-1})\Delta v(t+j)$$

Whereas the polynomial $E_j(z^{-1})\,D(z^{-1}) = H_j(z^{-1}) + z^{-j} H'_j(z^{-1})$, with $\delta(H_j(z^{-1})) = j-1$, the prediction equation can be rewritten as

$$\hat{y}(t+j|t) = G_j(z^{-1})\Delta u(t+j-1) + H_j(z^{-1})\Delta v(t+j) + G'_j(z^{-1})\Delta u(t-1) + H'_j(z^{-1})\Delta v(t)$$
$$+F_j(z^{-1})y(t) \tag{23}$$

Note that the last three terms of the right-hand side of this equation depend on past values of the process output, measured disturbances and input variables and correspond to the free response of the process considered if the control signals and measured disturbances are kept constant; while the first term only depends on future values of the control signals and can be interpreted as the force response, that is, the response obtained when the initial conditions are zero y(t-j) = 0, Δu(t-j-1) = 0, Δv(t-j) for j > 0.

The other terminus equation (23) depends on the future deterministic disturbance.

$$\hat{y}(t+j|t) = G_j(z^{-1})\Delta u(t+j-1) + H_j(z^{-1})\Delta v(t+j) + f_j$$

$$f_j = G'_j(z^{-1})\Delta u(t-1) + H'_j(z^{-1})\Delta v(t) + F_j(z^{-1})y(t)$$

Then for N j ahead predictions:

$$\hat{y}(t+N|t) = G_N(z^{-1})\Delta u(t+N-1) + H_j(z^{-1})\Delta v(t+N) + f_N$$

If one considers $H_j = \sum_{i=1}^{j} h_i z^{-1}$ where h_i are the coefficients of system step response to the disturbance, if f′ = Hv + f.

The predictive equation of is

$$Y = Gu + f'$$

5. Simulations results

5.1 Open loop

The following figures shows the behavior of each of the states against disturbances stair, rising and declining in each of the manipulated variables such as feed flow of the solution to concentrate, steam temperature, concentration of food and feed temperature. In each figure a, b, c and d correspond to 1, 2, 3 and 4 th respectively effect.

The following figure shows the response of the open loop system, when making a disturbance in one of the manipulated variables such as flow of food; in the figure 3 is represented the concentration of output in each of the effects and figure 4 is represented the temperature in each of the effects.

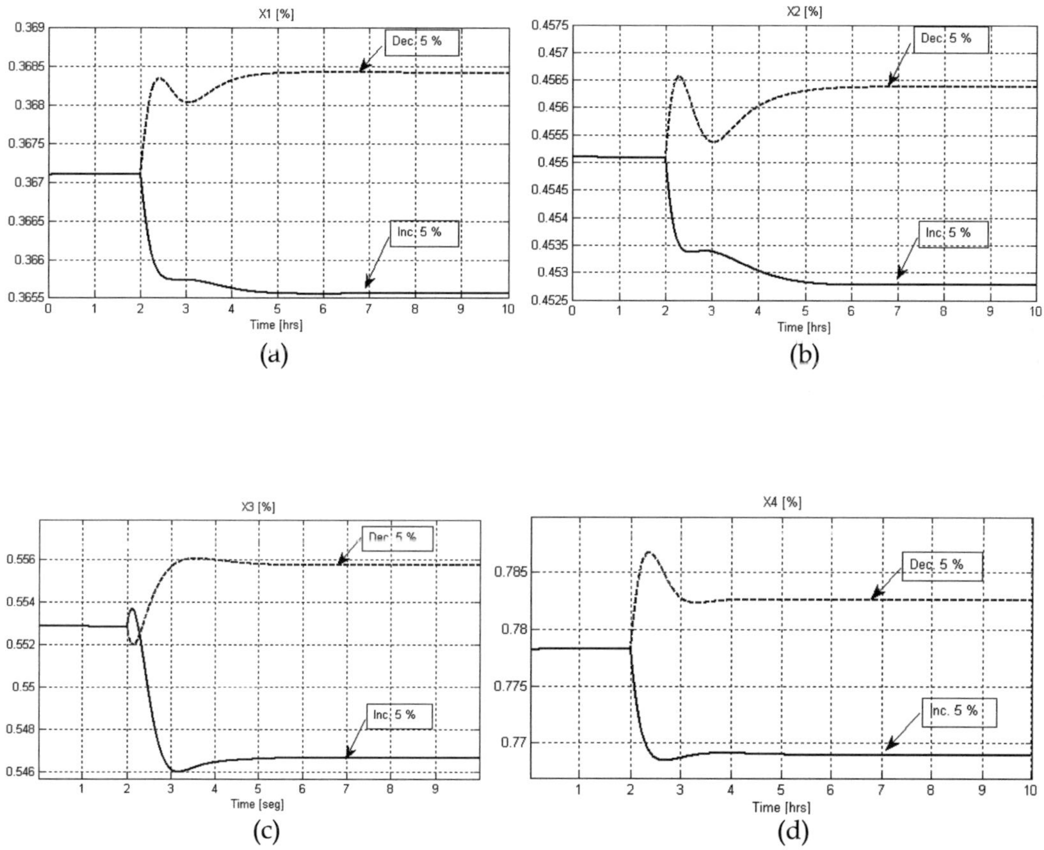

Fig. 3. Behavior of the outlet concentration of each of the effects of the evaporator to a change of a step in the flow of food (increase of 5% - decrease of 5%)

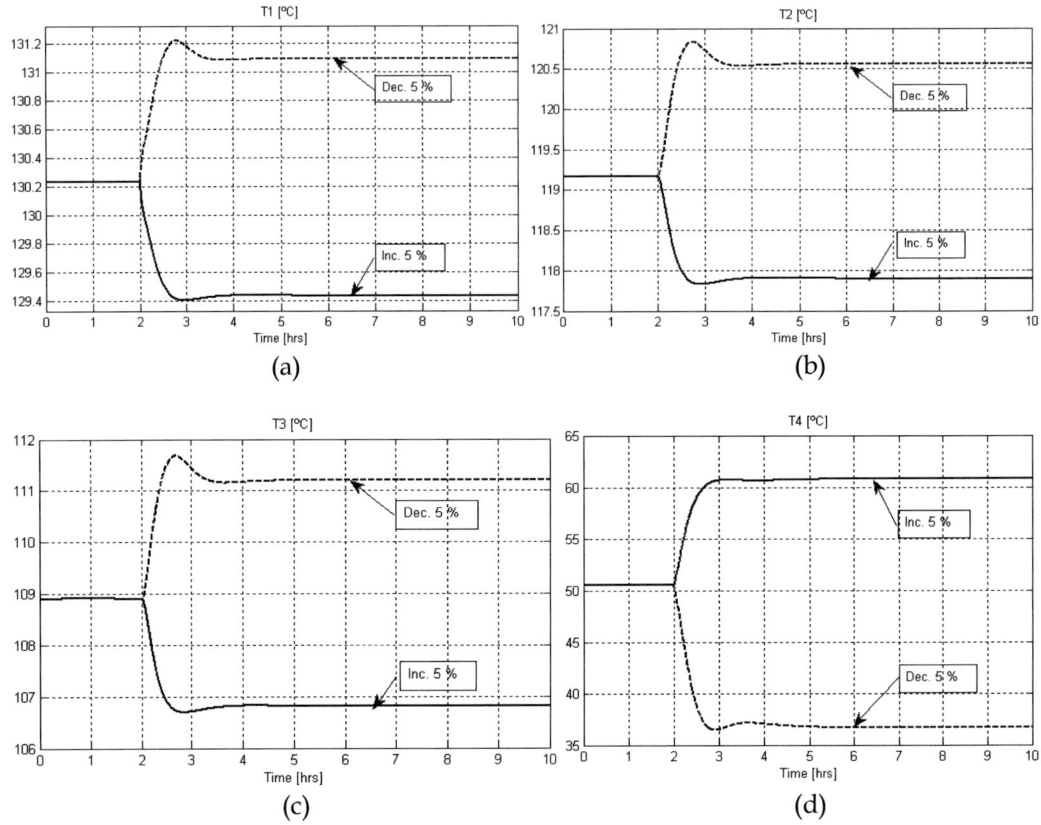

Fig. 4. Behavior of the temperature in the evaporator to a change of a step in the flow of food (increase of 5% - decrease of 5%)

In the following figures shows the response of the open loop system, when making a disturbance in one of the manipulated variables such as steam temperature is the other manipulated variable; in the figure 5 is represented the concentration of output in each of the effects and figure 6 is represented the temperature in each of the effects.

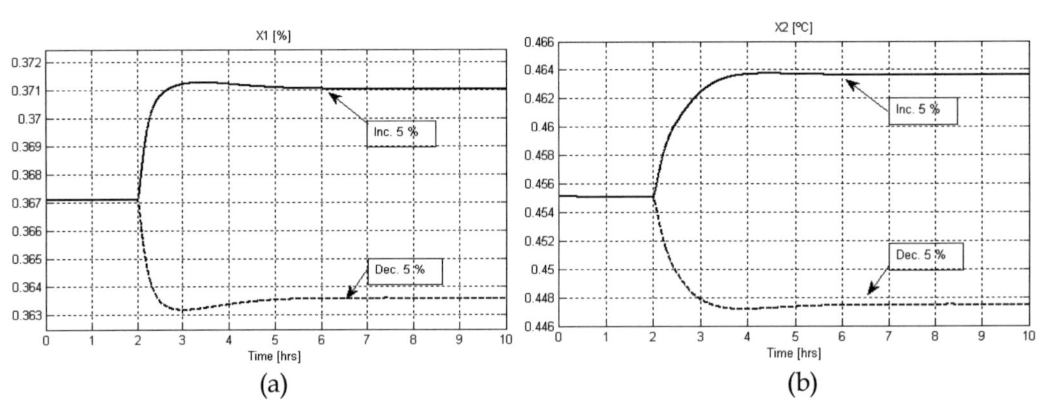

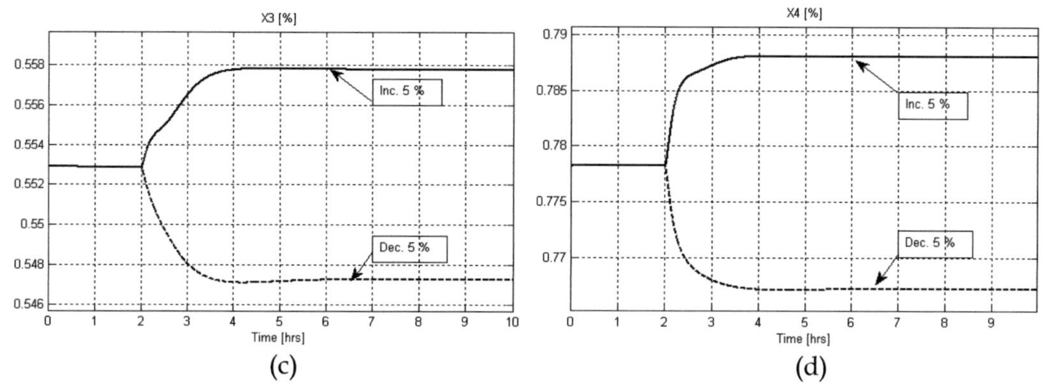

Fig. 5. Behavior of the concentration in the evaporator to a change of a step in the temperature of the steam supply (increase of 5% - decrease of 5%).

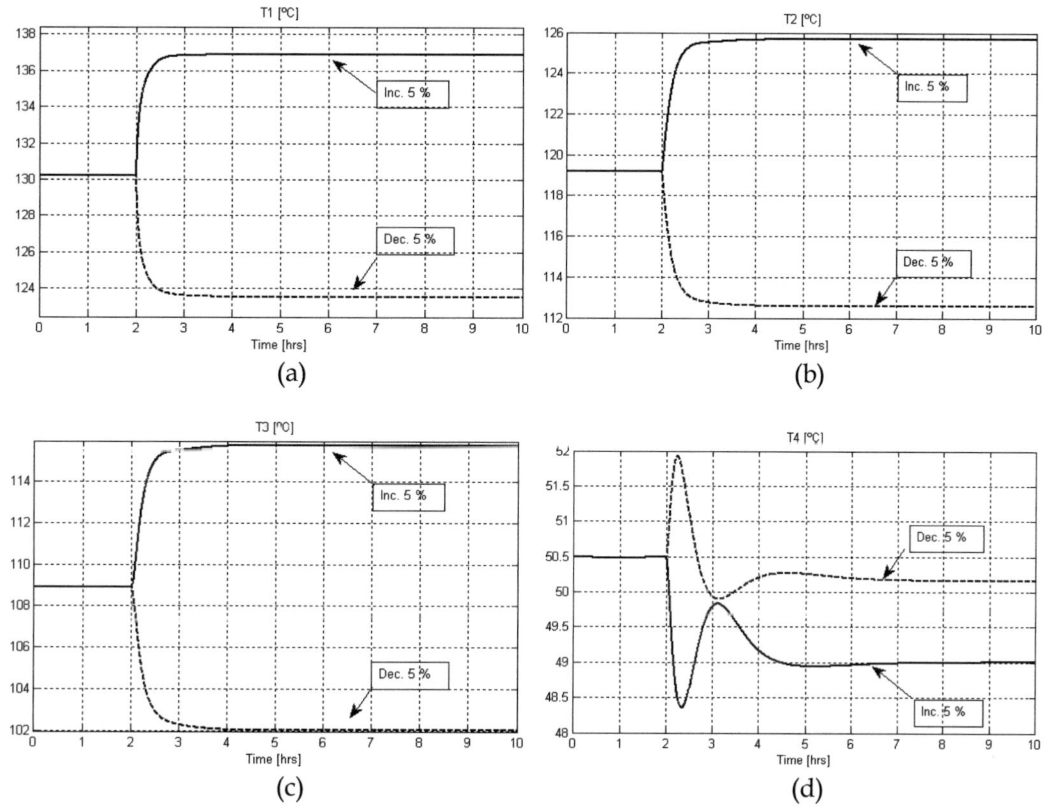

Fig. 6. Behavior of the temperature in the evaporator to a change of a step in the temperature of the steam supply (increase of 5% - decrease of 5%).

In the following figures shows the response of the open loop system, when making a step in one of the disturbance variables such as in feed concentration is one measurable disturbances; in the figure 7 is represented the concentration of output in each of the effects and figure 8 is represented the temperature in each of the effects.

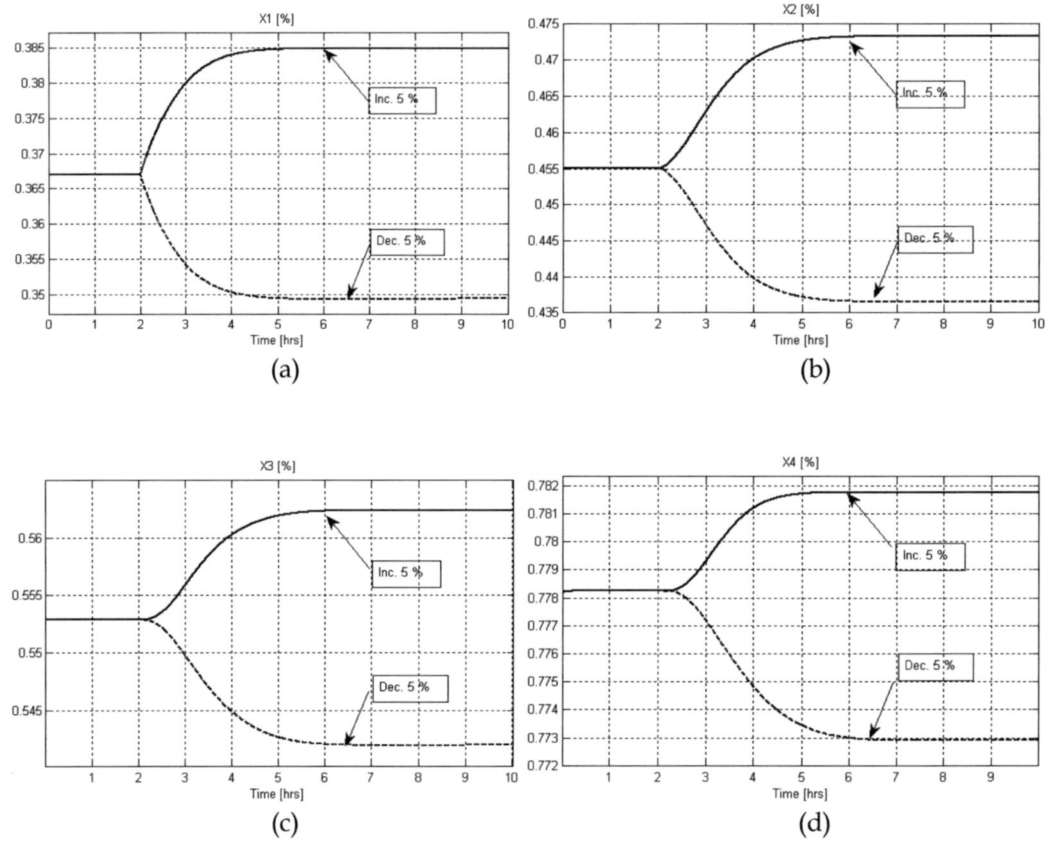

Fig. 7. Behavior of the concentration in the evaporator to a step change in feed concentration (increase of 5% - decrease of 5%).

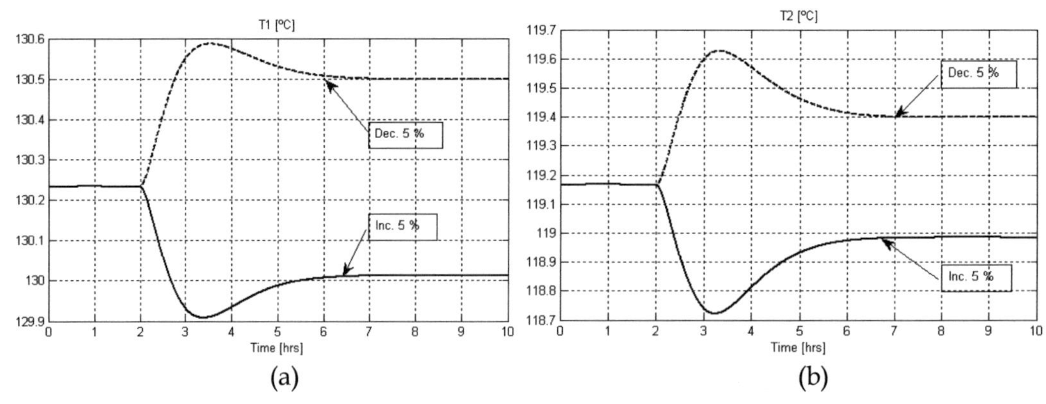

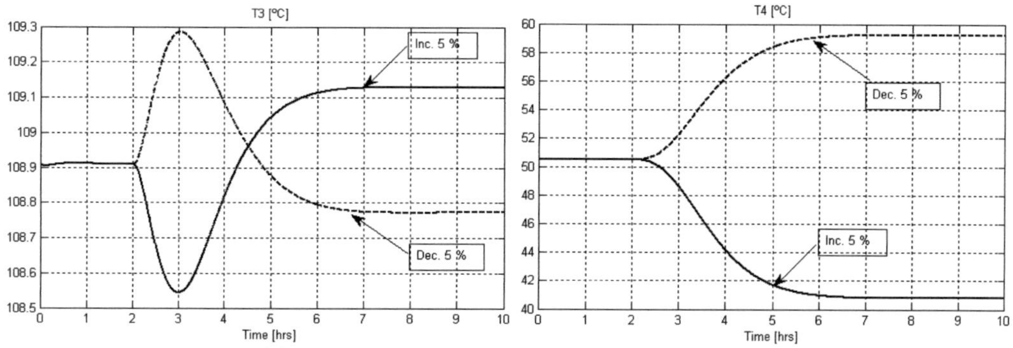

(c) (d)

Fig. 8. Behavior of the temperature in the evaporator to a step change in feed concentration (increase of 5% - decrease of 5%).

In the figures now shows the response of the open loop system, when making a disturbance in one of the disturbance variables such as in temperature of the input solution is the other measurable disturbances; in the figure 9 is represented the concentration of output in each of the effects and figure 10 is represented the temperature in each of the effects.

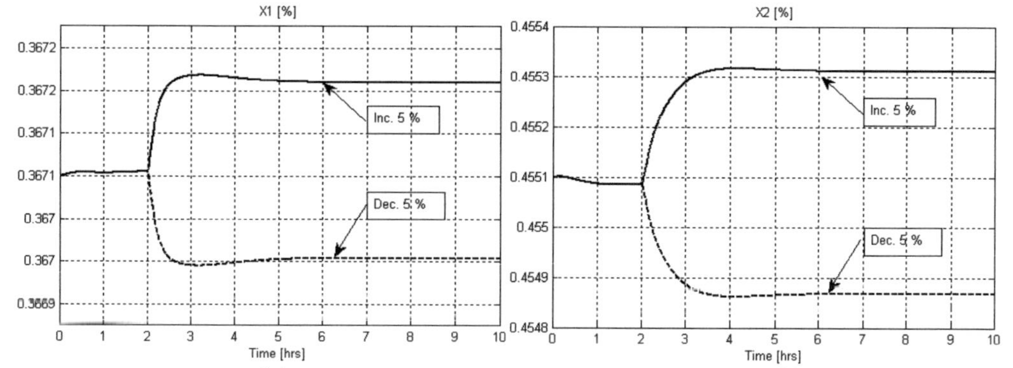

(a) (b)

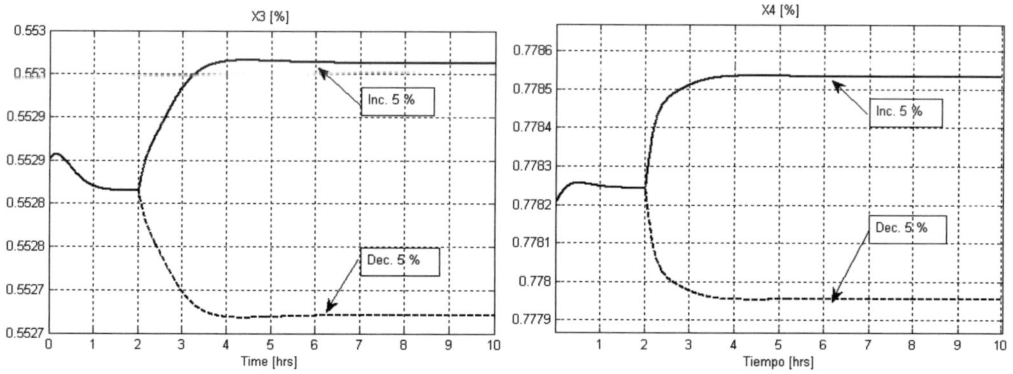

(c) (d)

Fig. 9. Behavior of the concentration in the evaporator to a change of a step in the temperature of the input solution (increase of 5% - decrease of 5%).

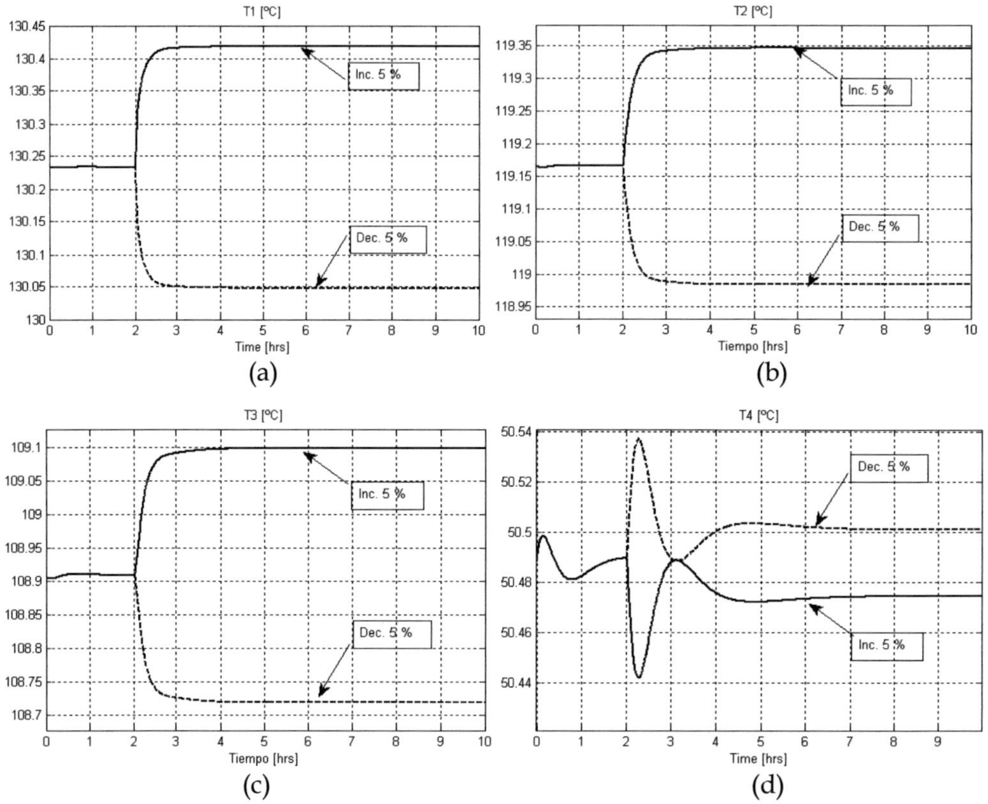

Fig. 10. Behavior of the temperature in the evaporator to a change of a step in the temperature of the input solution (increase of 5% - decrease of 5%).

5.2 Close loop

The following figures show the response of GPC controller, when conducted disturbances on the manipulated variables, ie giving an overview of the steam temperature and feed flow, one step at time 5 hours on the steam temperature and an increase to 10 hours in the feed stream.

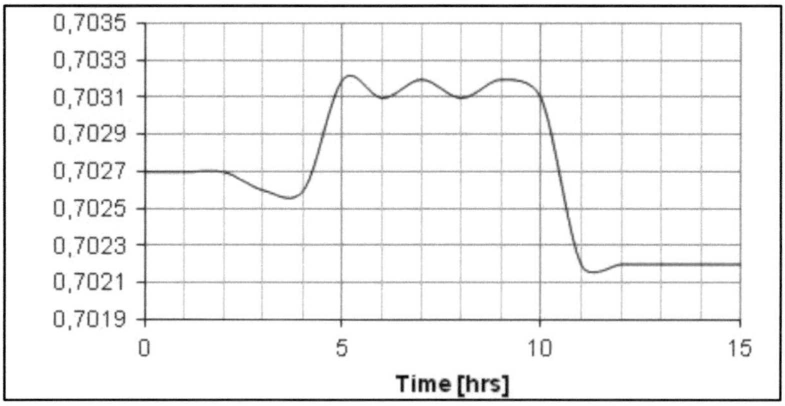

Fig. 11. Behavior of the final product concentration at the outlet of the fourth effect

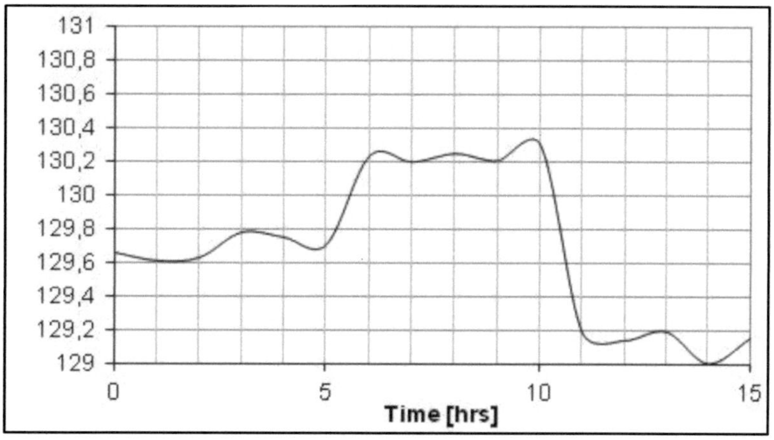

Fig. 12. Behavior of the temperature in the first effect

6. Conclusions

In analyzing the results obtained by performing perturbations in each of the four variables that enter the equipment, is considered appropriate the choice of manipulated variables chosen as the income flow of the solution to concentrate (grape juice) and the steam temperature and as measurable disturbances to the feed concentration and temperature that enters the solution concentration, this conclusion after observing emanates figures 3 to 10. We can also observe that the process of concentration has a complex dynamic, with long delays, high nonlinearity, coupling between variables, added to the reactions of deterioration of the organoleptic properties of the solution to concentrate

From the results shown in Figures 11 and 12 on the behavior of the controlled system verifies that the design of GPC has performed well since the variations in the controlled variable are smoother. As well as you can see the robustness of the proposed controller.

7. Acknowledgments

The authors gratefully acknowledge the financial support of the "Universidad de La Frontera"- Chile DIUFRO DI07-0102, "Universidad Nacional de San Juan"- Argentina, Project FI-I1018. They are also grateful for the cooperation of "Mostera Rio San Juan".

8. References

(Albertos, 1989) Albertos, P. and Ortega, R. "*On Generalized Predictive Control: Two Alternative Formulation*"*s*. Automatica, 25 (5): 753-755

(Allidina, 1980) Allidina A. Y. and Hughes, F. M. "*Generalized Self-tuning Controller with Pole Assignment*". Proccedings IEE, Part D, 127: 13-18.

(Armaou, 2002) Armaou A., Christofides P.D., "Dynamic Optimization of Dissipative PDE Systems Using Nonlinear Order Reduction". Chemical Engineering Science 57 - 24, pp. 5083-5114.

(Aros, 2008), Nelson Aros, Carlos Muñoz, Javier Mardones, Ernesto Villarroel, Ludwig Von Dossow, *"Sintonía de controladores PID basado en la respuesta de un GPC"*II Congreso Internacional de Ingeniería Electrónica, ciudad de Madero - México, marzo de 2008.

(Asghari, 2006) Asghari, F. S.; Yoshida, H. *"Acid-catalyzed production of 5-hydroxymethyl furfural from D-fructose in subcritical water"*. Ind. Eng. Chem. Res. 2006, 45, 2163–2173.

(Barbieri, 1980) Barbieri, R., & Rossi, N. "Propietà fisiche dei mosti d'uva concentrati". *Rivista de Viticol. e di Enologia. Conegliano,* No 1, 10–18.

(Bitmead, 1990) Bitmead, R.R.,M.Geversand V.Wertz *"Adaptive Optimal Control: the Thinking Man's GPC"*. Prentice-Hall.

(Camacho, 2004) Camacho, E.F. and Bordons, C. *"Model Predictive Control"*, Springer, London Limited.

(Clarke, 1987) Clarke, D.W. and Mohtadi, P.S. Tuffs, C. *" Generalized predictive control – Part I: thebasic algorithm"*, Automatica 23 (2) 137–148.

(Christofides, 1998) Christofides P.D, *"Robust Control of Parabolic PDE Systems"*, Chemical Engineering Science, 53, 16, 2949-2965.

(Di Leo, 1988) Di Leo, F. "Caratteristiche fisico-chimiche dei mosti concentrati rettificati. *Valutazione gleucometrica". Vignevini,* 15(1/2), 43–45.

(Ginz, 2000) Ginz, M.; Balzer, H. H.; Bradbury, A. G. W.; Maier, H. G. *"Formation of aliphatic acids by carbohydrate degradation during roasting of coffee"*. Eur. Food Res. Technol. 211, 404–410.

(Gogus, 1998) Gogus, F.; Bozkurt, H.; Eren, S. *"Kinetics of Maillard reaction between the major sugars and amino acids of boiled grape juice"*. Lebensm.-Wiss. Technol 31, 196–200.

(Ibarz, 1993) Ibarz, A., & Ortiz, J. "Reología de Zumos de Melocotón". Alimentación, Equipos y Tecnología. Octubre, 81–86, Instituto Nacional de Vitivinicultura. *Síntesis básica de estadística vitivinícola argentina, Mendoza.* Varios números.

(Kam, 1999) Kam K.M., Tade M.O., "Case studies on the modelling and control of evaporation systems". *XIX Interamerican Congress of Chemical Engineering COBEQ.*

(Kam, 2000) Kam K.M., Tade M.O., "Simulated Nonlinear Control Studies of Five Effect Evaporator Models". *Computers and Chemical Engineering, Vol. 23, pp. 1795 - 1810.*

(Doyle, 1979) Doyle J.C., Stein G., "Robustness with observers". *IEEE Trans. on Auto. Control, Vol. AC-24, April.*

(El-Farra, 2003) El-Farra N.H., Armaou A., Christofides P.D., "Analysis and Control of Parabolic PDE Systems with Input Constraints". *Automatica 39 – 4, pp. 715-725.*

(Freudenberg, 1988) Freudenberg J., Looze D., *Frequency Domain Properties of Scalar and Multivariable Feedback Systems.* Springer Verlag, Berlín.

(Friedland, 1989) Friedland B., "On the properties of reduced-orden Kalman filters". *IEEE Trans. on Auto. Control, Vol. AC-34, March.*

(Hoo,2001) Hoo, K.A. and D. Zheng, *"Low-Order Control Relevant Model for a Class of Distributed Parameter Systems"*, Chemical Engineering Science, 56, 23, 6683-6710.

(Kaya, 2002) Kaya A., Belibagh K.B., *"Rheology of solid Gaziantep Pekmez". Journal of Food Engineering, Vol. 54, pp. 221-226.*

(Lelic, 1987) Lelic, M. A. and Wellstead, P. E, *"Generalized Pole Placement Self Tuning Controller". Part 1"*. Basic Algorithm. International J. of Control, 46 (2): 547-568.

(Lelic, 1987) Lelic, M. A. and Zarrop, M. B. *"Generalized Pole Placement Self Tuning Controller. Part 2"*. Basic Algorithm Application to Robot Manipulator Control. International J. of Control, 46 (2): 569-601, 1987.

(Middleton, 1990) Middleton R.H., Goodwin G.C., *Digital Control and Estimation. A Unified Approach.* Prentice Hall, Englewood Cliffs, N.J.

(Moressi, 1984). Moressi, M., & Spinosi, M. "Engineering factors in the production of concentrated fruit juices, II, fluid physical properties of grapes". *Journal of Food Technology*, 5(19), 519–533.

(Ortiz, 2006) Ortiz, O.A., Suárez, G.I. & Mengual, C.A. "Evaluation of a neural model predictive control for the grape juice concentration process". XXII Congreso 2006.

(Pandolfi, 1991) Pandolfi, C., Romano, E. & Cerdán, A. Composición de los mostos concentrados producidos en Mendoza y San Juan, Argentina. Ed. Agro Latino. *Viticultura/Enología profesional* 13, 65–74.

(Perry, 1997) Perry R., *Perry's Chemical Engineers Handbook.* 7TH Edition McGraw Hill.

(Pilati, 1998) Pilati, M. A., Rubio, L. A., Muñoz, E., Carullo, C. A., Chernikoff, R.E. & Longhi, M. F. "Evaporadores tubulares de circulación forzada: consumo de potencia en distintas configuraciones. III Jornadas de Investigación´ n. FCAI - UNCuyo. Libro de Resúmenes, 40.

(Piva, 2008) Piva, A.; Di Mattia, C.; Neri, L.; Dimitri, G.; Chiarini, M.; Sacchetti, G. Heat-induced chemical, physical and functional changes during grape must cooking. Food Chem. 2008, 106 (3), 1057–1065.

(Rao, 1984) Rao, M. A., Cooley, H. J., & Vitali, A. A. "Flow properties of concentrated juices at low temperatures. *Food Technology*, 3(38), 113–119.

(Rubio, 1998) Rubio, L. A., Muñoz, E., Carullo, C. A., Chernikoff, R. E., Pilati, M. A. & Longhi, M. F. "Evaporadores tubulares de circulación forzada: capacidad de calor intercambiada en distintas configuraciones". *III Jornadas de Investigación. FCAI - UNCuyo. Libro de Resúmenes,* 40.

(Sáenz 1986) Sáenz, C., & Costell, E. "Comportamiento Reológico de Productos de Limón, Influencia de la Temperatura y de la Concentración". *Revista de Agroquímica y Tecnología de Alimentos,* 4(26), 581–588.

(Saravacos, 1970) Saravacos, G. D. "Effect of temperature on viscosity of fruit juices and purees". *Journal of Food Science,* (35), 122–125.

(Schwartz, 1986) Schwartz, M., & Costell, E. "Influencia de la Temperatura en el Comportamiento Reológico del Azúcar de Uva (cv, Thompson Seedless)". *Revista de Agroquímica y Tecnología de Alimentos,* 3(26), 365–372.

(Stefanov, 2003) Stefanov Z.I., Hoo K.A., "A Distributed-Parameter Model of Black Liquor Falling Film Evaporators". *Part I. Modeling of Single Plate. Industrial Engineering Chemical Research 42, 1925-1937.*

(Suarez, 2010) Suarez G.I., Ortiz O.A., Aballay P.M., Aros N.H., "Adaptive neural model predictive control for the grape juice concentration process". *International Conference on Industrial Technology, IEEE-ICIT 2010, Chile.*

(Zang, 1990) Zang Z., Freudenberg J.S., "Loop transfer recovery for nonminimum phase plants". *IEEE Trans. Automatic Control, Vol. 35, pp. 547-553.*

(Zheng, 2002) Zheng D., Hoo K. A., "Low-Order Model Identification for Implementable Control Solutions of Distributed Parameter Systems". *Computers and Chemical Engineering 26 7-8, pp. 1049-1076.*

(Zuritz, 2005) Zuritz C.A., Muñoz E., Mathey H.H., Pérez E.H., Gascón A., Rubio L.A., Carullo C.A., Chemikoff R.E., Cabeza M.S., *"Density, viscosity and soluble solid concentration and temperatures"*. Journal of Food Engineering, *Vol. 71, pp. 143 - 149.*

6

Nonlinear Model Predictive Control for Induction Motor Drive

Adel Merabet
Division of Engineering, Saint Mary's University, Halifax, NS, Canada

1. Introduction

The induction motor (IM) is widely used in industry because of its well known advantages such as simple construction, less maintenance, reliability and low cost. However, it is highly nonlinear, multivariable, time-varying system and, contrary to DC motor, requires more complex methods of control. Therefore, this machine constitutes a theoretically challenging control problem.

One of the most important development in control area for induction motor has been field oriented control (FOC) established firstly by (Blaschke, 1972). However, the performance of this technique is affected by the motor parameter variations and unknown external disturbances. To improve the dynamic response and reduce the complexity of FOC methods, an extension amount of work has been done to find new methods, such as direct torque control (DTC), sliding mode and nonlinear control (Barut et al., 2005; Chen & Dunnigan, 2003; Chiasson, 1996; Marino et al. 1993).

Model based predictive control (MPC) is one of the most promising control methods for both linear and nonlinear systems. The MPC formulation integrates optimal control, multivariable control, and the use of future references. It can also handle constraints and nonlinear processes, which are frequently found in industry. However, the computation of the MPC requires some mathematical complexities, and in the way of implementing and tuning this kind of controller, the computation time of the MPC may be excessive for the sampling time required by the process. Therefore, several MPC implementations were done for slow processes (Bordons & Camacho, 1998; Garica et al., 1989; Richalet, 1993). However, the explicit formulation of MPC allows its implementation in fast linear systems (Bemporad et al. 2002).

A review of fast method for implementing MPC can be found in (Camacho & Bordons, 2004). In case of nonlinear systems, where the mathematical packages are available in research control community, and thanks to the advancement of signal processing technology for control techniques, it becomes easy to implement these control schemes. Many works have been developed in nonlinear model predictive control (NMPC) theory (Ping, 1996; Chen et al., 1999; Siller-Alcala, 2001; Feng et al., 2002). A nonlinear PID model predictive controller developed in (Chen et al., 1999), for nonlinear control process, can improve some desirable features, such as, robustness to parameters variations and external disturbance rejection. The idea is to develop a nonlinear disturbance observer, and by embedding the nonlinear model predictive control law in the observer structure, it allows to express the disturbance observer through a PID control action. The NMPC have been implemented in induction motor drive with good performance (Hedjar et al., 2000; Hadjar et al. 2003; Maaziz et al., 2000; Merabet et al., 2006; Correa et al., 2007; Nemec et al., 2007). However, in these works, the load torque is taken as a known quantity to achieve accurately the desired performance, which is not always true in the majority of the industrial applications. Therefore, an observer for load torque is more than necessary for high performance drive. The design of such observer must not be complicated and well integrated in the control loop.

This chapter presents a nonlinear PID model predictive controller (NMPC PID) application to induction motor drive, where the load torque is considered as an unknown disturbance. A load torque observer is derived from the model predictive control law and integrated in the control strategy as PID speed controller. This strategy unlike other techniques for load torque observation (Marino et al., 1998; Marino et al., 2002; Hong & Nam, 1998; Du & Brdys, 1993), where the observer is an external part from the controller, allows integrating the observer into the model predictive controller to design a nonlinear PID model predictive controller, which improves the drive performance. It will be shown that the controller can be implemented with a limited set of computation and its integration in the closed loop scheme does not affect the system stability. In the development of the control scheme, it is assumed that all the machine states are measured. In fact a part of the state, the rotor flux, is not easily measurable and it is costly to use speed sensor. In literature, many techniques exist for state estimation (Jansen et al., 1994; Leonhard, 2001). A continuous nonlinear state observer based on the observation errors is used in this work to estimate the state variables. The coupling between the observer and the controller is analyzed, where the global stability of the whole system is proved using the Lyapunov stability. For this reason, a continuous version of NMPC is used in this work.

The rest of the chapter is organized as follows. In section 2, the induction motor model is defined by a nonlinear state space model. In section 3, the NMPC control law is developed for IM drive with an analysis of the closed loop system stability. In section 4, the load torque is considered as a disturbance variable in the machine model, and a NMPC PID control is applied to IM drive. Then, the coupling between the controller and the state observer is discussed in section 5, where the global stability of the whole system is proven theoretically. In section 6, simulation results are given to show the effectiveness of the proposed control strategy.

2. Induction motor modeling

The stator fixed (α-β) reference frame is chosen to represent the model of the motor. Under the assumption of linearity of the magnetic circuit, the nonlinear continuous time model of the IM is expressed as

$$\dot{\mathbf{x}}(t) = \mathbf{f}(\mathbf{x}) + \mathbf{g}_1(\mathbf{x})\mathbf{u}(t) \tag{1}$$

where

$$\mathbf{x} = \begin{bmatrix} i_{s\alpha} & i_{s\beta} & \phi_{r\alpha} & \phi_{r\beta} & \omega \end{bmatrix}^T, \ \mathbf{u} = \begin{bmatrix} u_{s\alpha} & u_{s\beta} \end{bmatrix}^T$$

The state **x** belongs to the set $\Omega = \left\{ \mathbf{x} \in \Re^5 : \phi_{r\alpha}^2 + \phi_{r\beta}^2 \neq 0 \right\}$.

Vector function **f(x)** and constant matrix $\mathbf{g_1(x)}$ are defined as follows.

$$\mathbf{f}(\mathbf{x}) = \begin{bmatrix} -\gamma i_{s\alpha} + \frac{K}{T_r}\phi_{r\alpha} + pK\omega\phi_{r\beta} \\ -\gamma i_{s\beta} + \frac{K}{T_r}\phi_{r\beta} - pK\omega\phi_{r\alpha} \\ \frac{L_m}{T_r} i_{s\alpha} - \frac{1}{T_r}\phi_{r\alpha} - p\omega\phi_{r\beta} \\ \frac{L_m}{T_r} i_{s\beta} - \frac{1}{T_r}\phi_{r\beta} + p\omega\phi_{r\alpha} \\ \frac{pL_m}{JL_r}\left(\phi_{r\alpha} i_{s\beta} - \phi_{r\beta} i_{s\alpha}\right) - \frac{f_r}{J}\omega - \frac{T_L}{J} \end{bmatrix} \qquad \mathbf{g_1} = \begin{bmatrix} g_{11} \, g_{12} \end{bmatrix} = \begin{bmatrix} \frac{1}{\sigma L_s} & 0 & 0 & 0 & 0 \\ 0 & \frac{1}{\sigma L_s} & 0 & 0 & 0 \end{bmatrix}^T$$

where

$$\sigma = 1 - \frac{L_m^2}{L_s L_r}; \; K = \frac{L_m}{\sigma L_s L_r}; \; \gamma = \frac{1}{\sigma L_s}\left(R_s + \frac{R_r L_m^2}{L_r^2} \right)$$

The outputs to be controlled are

$$\mathbf{y} = \mathbf{h}(\mathbf{x}) = \begin{bmatrix} \omega \\ \phi_r^2 = \phi_{r\alpha}^2 + \phi_{r\beta}^2 \end{bmatrix} \tag{2}$$

f(x) and **h(x)** are assumed to be continuously differentiable a sufficient number of time. $i_{s\alpha}$, $i_{s\beta}$ denote stator currents, $\phi_{r\alpha}$, $\phi_{r\beta}$ rotor fluxes, ω rotor speed, $u_{s\alpha}$, $u_{s\beta}$ stator voltages, R_s, R_r stator and rotor resistances, L_s, L_r, L_m stator, rotor and mutual inductances, p number of poles pair, J inertia of the machine, f_r friction coefficient, $T_r = L_r/R_r$ rotor time constant, σ leakage coefficient and T_L load torque.

3. Nonlinear model predictive control

Nonlinear model predictive control (NMPC) algorithm belongs to the family of optimal control strategies, where the cost function is defined over a future horizon

$$\Im(\mathbf{x},\mathbf{u}) = \frac{1}{2}\int_0^{\tau_r} \left(\mathbf{y}(t+\tau) - \mathbf{y_r}(t+\tau)\right)^T \left(\mathbf{y}(t+\tau) - \mathbf{y_r}(t+\tau)\right) d\tau \tag{3}$$

where τ_r is the prediction time, $\mathbf{y}(t+\tau)$ a τ-step ahead prediction of the system output and $\mathbf{y_r}(t+\tau)$ the future reference trajectory. The control weighting term is not included in the cost function (3). However, the control effort can be achieved by adjusting prediction time. More details about how to limit the control effort can be found in (Chen et al., 1999).

The objective of model predictive control is to compute the control $\mathbf{u}(t)$ in such a way the future plant output $\mathbf{y}(t+\tau)$ is driven close to $\mathbf{y_r}(t+\tau)$. This is accomplished by minimizing $\Im$.

The relative degree of the output, defined to be the number of times of output differentiation until the control input appears, is r_1=2 for speed output and r_2=2 for flux output. Taylor series expansion (5) can be used for the prediction of the machine outputs in the moving time frame. The differentiation of the outputs with respect to time is repeated r times.

$$y_i(t+\tau) = h_i(\mathbf{x}) + \tau L_{\mathbf{f}} h_i(\mathbf{x}) + \frac{\tau^2}{2!} L_{\mathbf{f}}^2 h_i(\mathbf{x}) + \ldots + \frac{\tau^{r_i}}{r_i!} L_{\mathbf{f}}^{r_i} h_i(\mathbf{x}) + \frac{\tau^{r_i}}{r_i!} L_{\mathbf{g}} L_{\mathbf{f}}^{(r_i-1)} h_i(\mathbf{x}) \mathbf{u}(t) \quad (4)$$

The predicted output $\mathbf{y}(t+\tau)$ is carried out from (4)

$$\mathbf{y}(t+\tau) = \overline{\mathbf{T}}(\tau)\mathbf{Y}(t) \quad (5)$$

where

$$\begin{cases} \overline{\mathbf{T}}(\tau) = \begin{bmatrix} I_{2\times2} & \tau I_{2\times2} & \tau^2/2\, I_{2\times2} \end{bmatrix} \\ I_{2\times2} : \text{Identity matrix} \end{cases}$$

The outputs differentiations are given in matrix form as

$$\mathbf{Y}(t) = \begin{bmatrix} \mathbf{y}(t) \\ \dot{\mathbf{y}}(t) \\ \ddot{\mathbf{y}}(t) \end{bmatrix} = \begin{bmatrix} \mathbf{h}(\mathbf{x}) \\ L_{\mathbf{f}}\mathbf{h}(\mathbf{x}) \\ L_{\mathbf{f}}^2\mathbf{h}(\mathbf{x}) \end{bmatrix} + \begin{bmatrix} 0_{2\times1} \\ 0_{2\times1} \\ \mathbf{G}_1(\mathbf{x})u(t) \end{bmatrix} \quad (6)$$

where

$$L_{\mathbf{f}}^i\mathbf{h}(\mathbf{x}) = \begin{bmatrix} L_{\mathbf{f}}^i h_1(\mathbf{x}) & L_{\mathbf{f}}^i h_2(\mathbf{x}) \end{bmatrix}^T, (i = 0,1,2)$$

$$\mathbf{G}_1(\mathbf{x}) = \begin{bmatrix} L_{g_{11}} L_{\mathbf{f}} h_1(\mathbf{x}) & L_{g_{12}} L_{\mathbf{f}} h_1(\mathbf{x}) \\ L_{g_{11}} L_{\mathbf{f}} h_2(\mathbf{x}) & L_{g_{12}} L_{\mathbf{f}} h_2(\mathbf{x}) \end{bmatrix} \quad (7)$$

A similar computation is used to find the predicted reference $\mathbf{y}_\mathbf{r}(t+\tau)$

$$\mathbf{y}_\mathbf{r}(t+\tau) = \overline{\mathbf{T}}(\tau)\mathbf{Y}_\mathbf{r}(t) \quad (8)$$

where

$$\begin{cases} \mathbf{Y}_\mathbf{r}(t) = \begin{bmatrix} \mathbf{y}_\mathbf{r}(t) & \dot{\mathbf{y}}_\mathbf{r}(t) & \ddot{\mathbf{y}}_\mathbf{r}(t) \end{bmatrix}^T \\ \mathbf{y}_\mathbf{r}(t) = \begin{bmatrix} \omega_{ref} & \phi_{ref}^2 \end{bmatrix}^T \end{cases}$$

Using (7) and (8), the cost function (3) can be simplified as

$$\Im(\mathbf{x},\mathbf{u}) = \frac{1}{2}\left(\mathbf{Y}(t) - \mathbf{Y}_\mathbf{r}(t)\right)^T \overline{\mathbf{\Pi}}\left(\mathbf{Y}(t) - \mathbf{Y}_\mathbf{r}(t)\right) \quad (9)$$

where

$$\bar{\mathbf{\Pi}} = \int_0^{\tau_r} \bar{\mathbf{T}}(\tau)^T \bar{\mathbf{T}}(\tau) d\tau = \left[\begin{array}{cc:c} \tau_r I_{2\times2} & \frac{\tau_r^2}{2} I_{2\times2} & \frac{\tau_r^3}{6} I_{2\times2} \\ \frac{\tau_r^2}{2} I_{2\times2} & \frac{\tau_r^3}{3} I_{2\times2} & \frac{\tau_r^4}{8} I_{2\times2} \\ \hdashline \frac{\tau_r^3}{6} I_{2\times2} & \frac{\tau_r^4}{8} I_{2\times2} & \frac{\tau_r^5}{20} I_{2\times2} \end{array}\right] = \begin{bmatrix} \bar{\mathbf{\Pi}}_1 & \bar{\mathbf{\Pi}}_2 \\ \bar{\mathbf{\Pi}}_2^T & \bar{\mathbf{\Pi}}_3 \end{bmatrix}$$

The optimal control is carried out by making $\partial\Im/\partial u = 0$

$$\mathbf{u}(t) = -\mathbf{G}_1(\mathbf{x})^{-1} [\bar{\mathbf{\Pi}}_3^{-1} \bar{\mathbf{\Pi}}_2^T \quad I_{2\times2}] \mathbf{M} \tag{10}$$

where

$$\begin{cases} \mathbf{M} = \begin{bmatrix} \mathbf{h}(\mathbf{x}) \\ L_{\mathbf{f}}\mathbf{h}(\mathbf{x}) \\ L_{\mathbf{f}}^2\mathbf{h}(\mathbf{x}) \end{bmatrix} - \begin{bmatrix} \mathbf{y_r}(t) \\ \dot{\mathbf{y}}_\mathbf{r}(t) \\ \ddot{\mathbf{y}}_\mathbf{r}(t) \end{bmatrix} \\ \det[\mathbf{G_1}(\mathbf{x})] = -\dfrac{2pL_m^2}{J\sigma^2 L_s^2 L_r T_r} \left(\phi_{r\alpha}^2 + \phi_{r\beta}^2\right) \end{cases}$$

The conditions $\{\phi_{r\alpha}(0), \phi_{r\beta}(0)\} \neq 0$ and the set $\Omega\{\phi_{r\alpha}^2 + \phi_{r\beta}^2 \neq 0\}$ allow $\mathbf{G_1}$ to be invertible. The singularity of this matrix occurs only at the start up of the motor, which can be avoided by putting initial conditions of the state observer different from zero. Let the optimal control (10) is developed as:

$$\mathbf{u}(t) = -\mathbf{G}_1(\mathbf{x})^{-1} \left(\sum_{i-0}^{2} \mathbf{K}_i \left(L_{\mathbf{f}}^i \mathbf{h}(\mathbf{x}) - \mathbf{y}_\mathbf{r}^{[i]}(t) \right) \right) \tag{11}$$

where

$$\mathbf{K}_0 = K_0 * I_{2\times2}; \mathbf{K}_1 = K_1 * I_{2\times2}; \mathbf{K}_2 = I_{2\times2}; \left(K_0 = {}^{10}\!/_{3\tau_r^2}; K_1 = {}^{5}\!/_{2\tau_r} \right)$$

4. Nonlinear PID predictive control

In the development of the NMPC, the load torque is taken as a known parameter and its values are used in the control law computation. In case, where the load torque is considered as an unknown disturbance, the nonlinear model of motor with the disturbance variable is given by

$$\dot{\mathbf{x}}(t) = \mathbf{f}(\mathbf{x}) + \mathbf{g}_1(\mathbf{x})\mathbf{u}(t) + \mathbf{g}_2(\mathbf{x})T_L(t) \tag{12}$$

where

$$\mathbf{g}_2 = [g_{21}] = \begin{bmatrix} 0 & 0 & 0 & 0 & -\frac{1}{J} \end{bmatrix}^T$$

The function $\mathbf{f}(\mathbf{x})$ in (12) is similar to the one in (1), but without the term $(-T_L/J)$.
We assume that the load torque follows this condition

$$\dot{T}_L(t) = 0 \tag{13}$$

Note that the assumption (13) does not necessarily mean a constant load torque, but that the changing rate of the load in every sampling interval should be far slower than the motor electromagnetic process. In reality this is often the case.

On the basis of equations (12), (13) and (9) it can be shown, in a manner similar to (10), that the optimal control becomes

$$\mathbf{u}(t) = -\mathbf{G}_1(\mathbf{x})^{-1}\left\{ [\bar{\mathbf{\Pi}}_3^{-1}\bar{\mathbf{\Pi}}_2^T \quad I_{2\times2}]\,\mathbf{M} + [\bar{\mathbf{\Pi}}_3^{-1}\bar{\mathbf{\Pi}}_2^T \quad I_{2\times2}]\,\mathbf{G}_2(\mathbf{x})T_L(t)\right\} \tag{14}$$

where

$$\mathbf{G}_2(\mathbf{x}) = \begin{bmatrix} 0 & 0 & L_{g_{21}}h_1(\mathbf{x}) & 0 & L_{g_{21}}L_{\mathbf{f}}h_1(\mathbf{x}) & 0 \end{bmatrix}^T$$

The optimal NMPC PID proposed in (Chen et al., 1999) has been developed for the same output and disturbance relative degrees. However, in the motor model (12), the disturbance relative degree is lower than the output one, which can be seen in the forms of $\mathbf{G}_1(\mathbf{x})$ and $\mathbf{G}_2(\mathbf{x})$. The same method is used in this work, to prove that even in this case a NMPC PID controller can be applied to induction motor drive.

From (12), we get

$$\mathbf{g}_2(\mathbf{x})T_L(t) = \dot{\mathbf{x}}(t) - \mathbf{f}(\mathbf{x}) - \mathbf{g}_1(\mathbf{x})\mathbf{u}(t) \tag{15}$$

An initial disturbance observer is given by

$$\dot{\hat{T}}_L(t) = -\mathbf{l}(\mathbf{x})\mathbf{g}_2(\mathbf{x})\hat{T}_L(t) + \mathbf{l}(\mathbf{x})\left\{\dot{\mathbf{x}}(t) - \mathbf{f}(\mathbf{x}) - \mathbf{g}_1(\mathbf{x})\mathbf{u}(t)\right\} \tag{16}$$

In (16), $\mathbf{l}(x) \in \Re^5$ is a gain vector to be designed.
The error of the disturbance observer is

$$e_{T_L}(t) = T_L(t) - \hat{T}_L(t) \tag{17}$$

Then, the error dynamic is governed by

$$\dot{e}_{T_L}(t) + \mathbf{l}(x)\mathbf{g}_2(x)e_{T_L}(t) = 0 \tag{18}$$

It can be shown that the observer is exponentially stable when

$$\mathbf{l}(\mathbf{x})\mathbf{g}_2(\mathbf{x}) = c, \quad c > 0 \tag{19}$$

The disturbance (load torque) T_L is replaced by its estimated value in the control law given by (14); which then becomes

$$\mathbf{u}(t) = -\mathbf{G}_1(\mathbf{x})^{-1}\left\{ [\bar{\mathbf{\Pi}}_3^{-1}\bar{\mathbf{\Pi}}_2^T \quad I_{2\times 2}]\,\mathbf{M} + [\bar{\mathbf{\Pi}}_3^{-1}\bar{\mathbf{\Pi}}_2^T \quad I_{2\times 2}]\,\mathbf{G}_2(\mathbf{x})\hat{T}_L(t)\right\} \tag{20}$$

Substituting (20) into (16) yields

$$\begin{aligned}\dot{\hat{T}}_L &= \mathbf{l}(\dot{\mathbf{x}} - \mathbf{f}) - \mathbf{l}\mathbf{g}_2\hat{T}_L - \mathbf{l}\mathbf{g}_1\mathbf{u} \\ &= \mathbf{l}(\dot{\mathbf{x}} - \mathbf{f}) - \mathbf{l}\mathbf{g}_2\hat{T}_L + \mathbf{l}\mathbf{g}_1\left(\mathbf{G}_1^{-1}\left\{ [\bar{\mathbf{\Pi}}_3^{-1}\bar{\mathbf{\Pi}}_2^T \quad I_{2\times 2}]\mathbf{M} + [\bar{\mathbf{\Pi}}_3^{-1}\bar{\mathbf{\Pi}}_2^T \quad I_{2\times 2}]\mathbf{G}_2\hat{T}_L]\right\}\right)\end{aligned} \tag{21}$$

Based on the definition of $\mathbf{G}_2(\mathbf{x})$, (14) and the condition (19), let's define (see B6)

$$\mathbf{l}(\mathbf{x}) = p_0\left(\frac{\partial L_{\mathbf{f}} h_1(\mathbf{x})}{\partial \mathbf{x}} + K_1\frac{\partial h_1}{\partial x}\right), \quad p_0 \neq 0 \text{ is a constant} \tag{22}$$

Substituting $\mathbf{l}(\mathbf{x})$ into (21), and using Lie derivatives simplifications (see appendix B), we get a simple form for load torque disturbance estimator.

$$\dot{\hat{T}}_L = p_0\left\{(\ddot{\omega} - \ddot{\omega}_{ref}) + K_1(\dot{\omega} - \dot{\omega}_{ref}) + K_0(\omega - \omega_{ref})\right\} \tag{23}$$

Integrating (23), we get

$$\hat{T}_L = p_0\left(\dot{e}_\omega(t) + K_1 e_\omega(t) + K_0\int_0^t e_\omega(\tau)d\tau\right) \tag{24}$$

The structure of this observer is driven by three tunable parameters, where p_0 is an independent parameter and K_i (i=0, 1) depend on the controller prediction horizon τ_r. It can be seen that the load toque observer has a PID structure, where the information needed is the speed error. Compared to the work in (Marino et al., 1993), where the load torque is estimated only via speed error, the disturbance observer (24) contains an integral action, which allows the elimination of the steady state error and enhances the robustness of the control scheme with respect to model uncertainties and disturbances rejection.

5. Global stability analysis

Initially, the model predictive control law is carried out assuming all the states are known by measurement, which is not always true in the majority of industrial applications. In fact, the rotor flux is not easily measurable. Therefore, a state observer must be used to estimate it. However, the coupling between the nonlinear model predictive control and the observer must guarantee the global stability.

5.1 Nonlinear state observer

To estimate the state, several methods are possible such as the observers using the observation errors for correction, which are powerful and improve the results. To construct an observer for the induction motor, written in (α, β) frame, the measurements of the stator voltages and currents are used in the design.

The real state, estimated state and observation errors are

$$\begin{cases} \mathbf{x} = \left[i_{s\alpha}\ i_{s\beta}\ \phi_{r\alpha}\ \phi_{r\beta}\ \omega \right]^T \\ \hat{\mathbf{x}} = \left[\hat{i}_{s\alpha}\ \hat{i}_{s\beta}\ \hat{\phi}_{r\alpha}\ \hat{\phi}_{r\beta}\ \hat{\omega} \right]^T \\ \tilde{\mathbf{x}} = \mathbf{x} - \hat{\mathbf{x}} \end{cases} \quad (25)$$

The state observer, derived from the motor model (1) with stator current errors for correction, is defined by

$$\dot{\hat{\mathbf{x}}} = \begin{bmatrix} -\gamma \hat{i}_{s\alpha} + \frac{K}{T_r}\hat{\phi}_{r\alpha} + pK\hat{\omega}\,\hat{\phi}_{r\beta} \\ -\gamma \hat{i}_{s\beta} + \frac{K}{T_r}\hat{\phi}_{r\beta} - pK\hat{\omega}\,\hat{\phi}_{r\alpha} \\ \frac{L_m}{T_r}\hat{i}_{s\alpha} - \frac{1}{T_r}\hat{\phi}_{r\alpha} - p\hat{\omega}\,\hat{\phi}_{r\beta} \\ \frac{L_m}{T_r}\hat{i}_{s\beta} - \frac{1}{T_r}\hat{\phi}_{r\beta} + p\hat{\omega}\,\hat{\phi}_{r\alpha} \\ \frac{pL_m}{JL_r}\left(\hat{\phi}_{r\alpha}\hat{i}_{s\beta} - \hat{\phi}_{r\beta}\hat{i}_{s\alpha}\right) - \frac{f_r}{J}\hat{\omega} - \frac{1}{J}T_L \end{bmatrix} + \begin{bmatrix} \frac{1}{\sigma L_s} & 0 \\ 0 & \frac{1}{\sigma L_s} \\ 0 & 0 \\ 0 & 0 \\ 0 & 0 \end{bmatrix}\mathbf{u} + \begin{bmatrix} k_1 & 0 \\ 0 & k_1 \\ \frac{k_2}{T_r} & -p\hat{\omega}\,k_2 \\ p\hat{\omega}\,k_2 & \frac{k_2}{T_r} \\ k_3 & k_3 \end{bmatrix}\begin{bmatrix} \tilde{i}_{s\alpha} \\ \tilde{i}_{s\beta} \end{bmatrix} + \begin{bmatrix} f_{ia} \\ f_{ib} \\ 0 \\ 0 \\ 0 \end{bmatrix} \quad (26)$$

$T_L = \hat{T}_L + e_{T_L}(t)$ and (f_{ia}, f_{ib}) are additional terms added in the observer structure, in order to establish the global stability of the whole system.

5.2 Control scheme based on state observer

The process states are used in the predictive control law design. However, in case of the IM, the states are estimated by (26). Including this observer in the control scheme allows defining the outputs (2) by

$$\begin{cases} \hat{h}_1 = \hat{\omega} \\ \hat{h}_2 = \hat{\phi}_{r\alpha}^2 + \hat{\phi}_{r\beta}^2 \end{cases} \quad (27)$$

The relative degrees are r_1=2 and r_2=2. Then, the first Lie derivatives of $\hat{h}_1$ and $\hat{h}_2$ are obtained by

$$\begin{cases} \dot{\hat{h}}_1 = L_{\hat{\mathbf{f}}}\hat{h}_1 \\ \dot{\hat{h}}_2 = L_{\hat{\mathbf{f}}}\hat{h}_2 \end{cases} \quad (28)$$

In (28), $\hat{\mathbf{f}}$ is the function of the motor model expressed with estimated states. Since $\dot{\hat{h}}_1$ and $\dot{\hat{h}}_2$ are not functions of the control inputs, one should derive them once again. However, they contain terms which are functions of currents. The differentiation of those terms introduces terms of flux, which are unknown. To overcome this problem, auxiliary outputs are introduced (Chenafa et al., 2005; Van Raumer, 1994) as

$$\begin{cases} L_{\hat{f}}\hat{h}_1 = \hat{h}_{11} - \dfrac{f_r}{J}\hat{h}_1 - \dfrac{T_L}{J} \\ L_{\hat{f}}\hat{h}_2 = -\dfrac{2}{T_r}\hat{h}_2 + \hat{h}_{21} + \Delta \end{cases} \quad (29)$$

where

$$\hat{h}_{11} = \frac{pL_m}{JL_r}(\hat{\phi}_{r\alpha}\hat{i}_{s\beta} - \hat{\phi}_{r\beta}\hat{i}_{s\alpha})$$

$$\hat{h}_{21} = \frac{2L_m}{T_r}(\hat{\phi}_{r\alpha}\hat{i}_{s\alpha} + \hat{\phi}_{r\beta}\hat{i}_{s\beta})$$

$$\Delta = 2\left(\frac{k_2}{T_r}\hat{\phi}_{r\alpha} + k_2 p\hat{\omega}\hat{\phi}_{r\beta}\right)\tilde{i}_{s\alpha} + 2\left(\frac{k_2}{T_r}\hat{\phi}_{r\beta} - k_2 p\hat{\omega}\hat{\phi}_{r\alpha}\right)\tilde{i}_{s\beta}$$

The derivatives of $\hat{h}_{11}$ and $\hat{h}_{21}$ are given by

$$\begin{cases} \dot{\hat{h}}_{11} = L_{\hat{f}}\hat{h}_{11} + L_{g_{11}}\hat{h}_{11}u_{s\alpha} + L_{g_{12}}\hat{h}_{11}u_{s\beta} \\ \dot{\hat{h}}_{21} = L_{\hat{f}}\hat{h}_{21} + L_{g_{11}}\hat{h}_{21}u_{s\alpha} + L_{g_{12}}\hat{h}_{21}u_{s\beta} \end{cases} \quad (30)$$

where

$$L_{\hat{f}}\hat{h}_{11} = f(\hat{i}_{s\alpha}, \hat{i}_{s\beta}, \hat{\phi}_{r\alpha}, \hat{\phi}_{r\beta}, i_{s\alpha}, i_{s\beta}, \omega);$$

$$L_{g_{11}}\hat{h}_{11} = -\frac{pL_m}{J\sigma L_s L_r}\hat{\phi}_{r\beta}; \quad L_{g_{12}}\hat{h}_{11} = \frac{pL_m}{J\sigma L_s L_r}\hat{\phi}_{r\alpha}$$

$$L_{\hat{f}}\hat{h}_{21} = f(\hat{i}_{s\alpha}, \hat{i}_{s\beta}, \hat{\phi}_{r\alpha}, \hat{\phi}_{r\beta}, i_{s\alpha}, i_{s\beta}, \omega);$$

$$L_{g_{11}}\hat{h}_{21} = \frac{2L_m}{\sigma L_s T_r}\hat{\phi}_{r\alpha}; \quad L_{g_{12}}\hat{h}_{21} = \frac{2L_m}{\sigma L_s T_r}\hat{\phi}_{r\beta}$$

This leads to

$$\begin{bmatrix} \dot{\hat{h}}_1 \\ \dot{\hat{h}}_{11} \\ \dot{\hat{h}}_2 \\ \dot{\hat{h}}_{21} \end{bmatrix} = \begin{bmatrix} \hat{h}_{11} - \dfrac{f_r}{J}\hat{h}_1 - \dfrac{T_L}{J} \\ L_{\hat{f}}\hat{h}_{11} + L_{g_{11}}\hat{h}_{11}u_{s\alpha} + L_{g_{12}}\hat{h}_{11}u_{s\beta} \\ -\dfrac{2}{T_r}\hat{h}_2 + \hat{h}_{21} + \Delta \\ L_{\hat{f}}\hat{h}_{21} + L_{g_{11}}\hat{h}_{21}u_{s\alpha} + L_{g_{12}}\hat{h}_{21}u_{s\beta} \end{bmatrix} \quad (31)$$

The errors between the desired trajectories of the outputs and the estimated outputs are

$$\begin{cases} e_1 = \hat{h}_1 - h_{1r} \\ e_2 = \hat{h}_{11} - h_{11r} \\ e_3 = \hat{h}_2 - h_{2r} \\ e_4 = \hat{h}_{21} - h_{21r} \end{cases} \tag{32}$$

Using (31), (32), the estimated states and the auxiliary outputs, the predictive control law (11), developed above through the cost function (3) minimization, becomes

$$\begin{bmatrix} u_{s\alpha} \\ u_{s\beta} \end{bmatrix} = \begin{bmatrix} L_{g_{11}}\hat{h}_{11} & L_{g_{12}}\hat{h}_{11} \\ L_{g_{11}}\hat{h}_{21} & L_{g_{12}}\hat{h}_{21} \end{bmatrix}^{-1} \begin{bmatrix} -L_{\hat{\mathbf{f}}}\hat{h}_{11} - e_1 - K_1 e_2 + \dot{h}_{11r} \\ -L_{\hat{\mathbf{f}}}\hat{h}_{21} - e_3 - K_1 e_4 + \dot{h}_{21r} \end{bmatrix} \tag{33}$$

The decoupling matrix in (33) is the same as in (7), since $L_{g_{1i}}\hat{h}_{11} = L_{g_{1i}}L_{\hat{\mathbf{f}}}\hat{h}_1$ and $L_{g_{1i}}\hat{h}_{21} = L_{g_{1i}}L_{\hat{\mathbf{f}}}\hat{h}_2;\ i = 1,\ 2$

From (31), (32) and (33), we get the error dynamic as

$$\begin{bmatrix} \dot{e}_1 \\ \dot{e}_2 \\ \dot{e}_3 \\ \dot{e}_4 \end{bmatrix} = \begin{bmatrix} \hat{h}_{11} - \frac{f_r}{J}\hat{h}_1 - \frac{T_L}{J} - \dot{h}_{1r} \\ -K_1 e_2 - e_1 \\ -\frac{2}{T_r}\hat{h}_2 + \hat{h}_{21} + \Delta - \dot{h}_{2r} \\ -K_1 e_4 - e_3 \end{bmatrix} \tag{34}$$

The references h_{1r} and h_{2r} and their derivatives are considered known.

In order to have (34) under the form given in (35) below, to use it in Lyapunov candidate, the references h_{11r} and h_{21r} must be defined as in (36)

$$\begin{bmatrix} \dot{e}_1 \\ \dot{e}_2 \\ \dot{e}_3 \\ \dot{e}_4 \end{bmatrix} = \begin{bmatrix} -K_0 e_1 + e_2 \\ -K_1 e_2 - e_1 \\ -K_0 e_3 + e_4 + \Delta \\ -K_1 e_4 - e_3 \end{bmatrix} \tag{35}$$

$$\begin{cases} h_{11r} = \frac{f_r}{J}\hat{h}_1 + \dot{h}_{1r} - K_0 e_1 + \frac{T_L}{J} \\ h_{21r} = \frac{2}{T_r}\hat{h}_2 + \dot{h}_{2r} - K_0 e_3 \end{cases} \tag{36}$$

An appropriate choice of K_0, K_1 ensures the exponential convergence of the tracking errors.

We now consider all the elements together in order to build a nonlinear model predictive control law based on state observer.

The functions V_1 and V_2, given by (37) and (38) below, are chosen to create a Lyapunov function candidate for the entire system (process, observer and controller); where γ_2 is a positive constant.

$$V_1 = \frac{\tilde{i}_{s\alpha}^2 + \tilde{i}_{s\beta}^2}{2} + \frac{\tilde{\phi}_{r\alpha}^2 + \tilde{\phi}_{r\beta}^2}{2\gamma_2} \tag{37}$$

$$V_2 = \frac{e_1^2 + e_2^2 + e_3^2 + e_4^2 + e_5^2}{2} \tag{38}$$

where, $e_5 = e_{T_L}$, represents the load torque observation error driven by the equation (18).

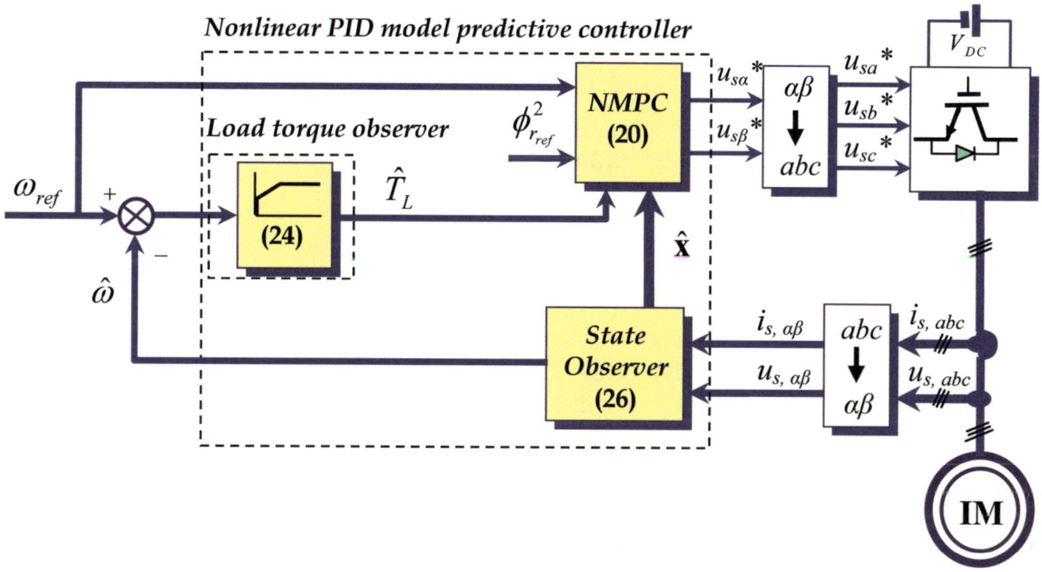

Fig. 1. Block diagram of the proposed nonlinear predictive sensorless control system.

The Lyapunov function and its derivative are respectively

$$V = V_1 + V_2 \tag{39}$$

$$\begin{aligned}\dot{V} = &\left(-K_0 e_1^2 - K_1 e_2^2 - K_0 e_3^2 - K_1 e_4^2 - c e_5^2\right) - (\gamma + k_1)\left(\tilde{i}_{s\alpha}^2 + \tilde{i}_{s\beta}^2\right) - \frac{1}{T_r\gamma_2}\left(\tilde{\phi}_{r\alpha}^2 + \tilde{\phi}_{r\beta}^2\right) + \\ &p\hat{\omega}\left(K - \frac{k_2}{\gamma_2}\right)\left(\tilde{i}_{s\alpha}\tilde{\phi}_{r\beta} - \tilde{i}_{s\beta}\tilde{\phi}_{r\alpha}\right) + \left(\frac{K}{T_r} + \frac{L_m}{T_r\gamma_2} - \frac{k_2}{T_r\gamma_2}\right)\left(\tilde{i}_{s\alpha}\tilde{\phi}_{r\alpha} + \tilde{i}_{s\beta}\tilde{\phi}_{r\beta}\right) - \left(f_{ia}\tilde{i}_{s\alpha} + f_{ib}\tilde{i}_{s\beta}\right) + \Delta e_3\end{aligned} \tag{40}$$

The following conditions form a sufficient set ensuring $\dot{V} < 0$

$$\begin{cases} k_2 = K\gamma_2 \\ -[f_{ia}\tilde{i}_{s\alpha} + f_{ib}\tilde{i}_{s\beta}] + \Delta e_3 = 0 \end{cases} \tag{41}$$

Replacing Δ by its value leads to the following equation

$$\left[f_{ia}\tilde{i}_{s\alpha}+f_{ib}\tilde{i}_{s\beta}\right]=2\left(\frac{k_2}{T_r}\hat{\phi}_{r\alpha}+k_2p\hat{\omega}\hat{\phi}_{r\beta}\right)\tilde{i}_{s\alpha}e_3+2\left(\frac{k_2}{T_r}\hat{\phi}_{r\beta}-k_2p\hat{\omega}\hat{\phi}_{r\alpha}\right)\tilde{i}_{s\beta}e_3 \tag{42}$$

Equation (42) is satisfied if f_{ia} and f_{ib} are chosen as

$$\begin{cases} f_{ia}=2\left(\frac{k_2}{T_r}\hat{\phi}_{r\alpha}+k_2p\hat{\omega}\hat{\phi}_{r\beta}\right)e_3 \\ f_{ib}=2\left(\frac{k_2}{T_r}\hat{\phi}_{r\beta}-k_2p\hat{\omega}\hat{\phi}_{r\alpha}\right)e_3 \end{cases} \tag{43}$$

V is then a Lyapunov candidate function for the overall system, formed by the process, the observer and the controller. Hence, the whole process is stable and the convergence is exponential.

6. Simulation results and discussion

In order to test all cases of IM operations, smooth references are taken for reversal speed and low speed. The results are compared with those of the standard FOC controller. The load torque disturbance is estimated by the observer (24) discussed above, which is combined with NMPC to create NMPC PID controller. The 1.1 kW induction motor (appendix D), which is fed by a SVPWM inverter switching frequency of 10 kHz, run with a sample time of 10 μs. The voltage input is given from the controller at the sample time T_s = 100 μs. The tuning parameters are the prediction time τ_r, the disturbance observer gain p_0 and (k_1, k_2, k_3) the gains of the state observer. All parameters are chosen by trial and error in order to achieve a successful tracking performance. The most important are (τ_r= 10*T_s, p_0=-0.001), which are used in all tests.

Figures 2 and 3 present the results for rotor speed and rotor flux norm tracking responses for the NMPC PID controller and for the well-known Field Oriented Controller (FOC). Figure 4 shows the components of the stator voltage and current. It can be seen that the choice of the prediction time τ_r has satisfied the tracking performance and the constraints on the signal control to be inside the saturation limits. Figure 5 gives the estimated load torque for different conditions of speed reference in the case of the proposed controller. As shown, the tracking performance is satisfactory achieved and the effect of the load torque disturbance on the speed is rapidly eliminated compared with the FOC strategy. Figures 6 to 8 present the proposed NMPC PID tracking performances for low speed operation. These results are also compared to those obtained by the FOC. As shown, the tracking performance is satisfactory achieved even at low speed.

In order to check the sensitivity of the controller and the state observer with respect to the parametric variations of the machine, these parameters are varied as shown in figure 9. It is to be noted that the motor model is affected by these variations, while the controller and the state observer are carried out with the nominal values of the machine parameters. The same values of tunable parameters (τ_r, p_0, k_1, k_2, k_3) have been used to show the influence of the parameters variations on the controller performance.

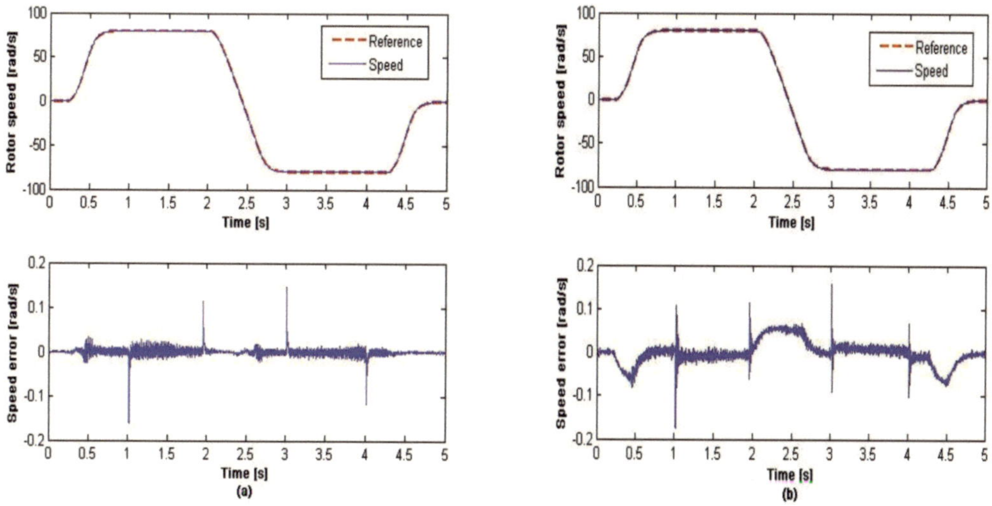

Fig. 2. Speed tracking performances - (a) proposed NMPC PID Controller, and (b) Field Oriented Controller (FOC).

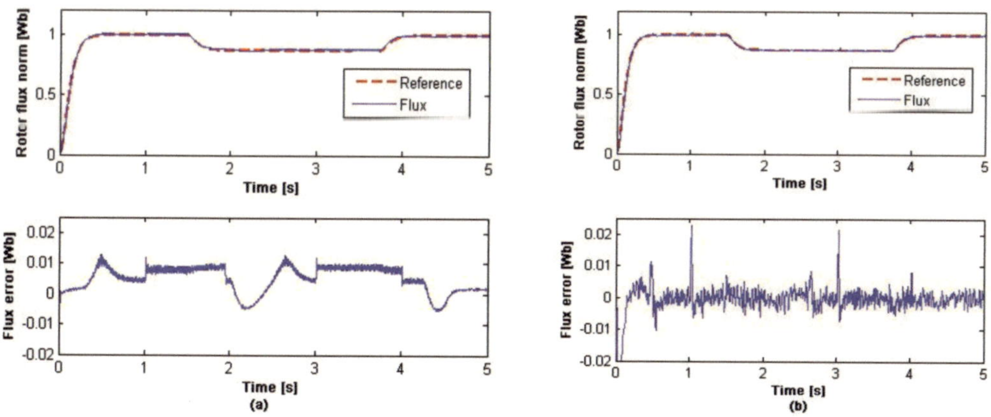

Fig. 3. Flux norm tracking performances - (a) proposed NMPC PID Controller, and (b) Field Oriented Controller (FOC).

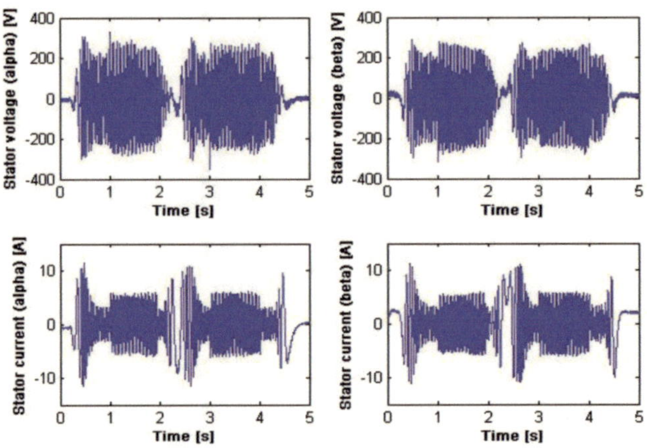

Fig. 4. Stator voltage and current components with NMPC PID controller

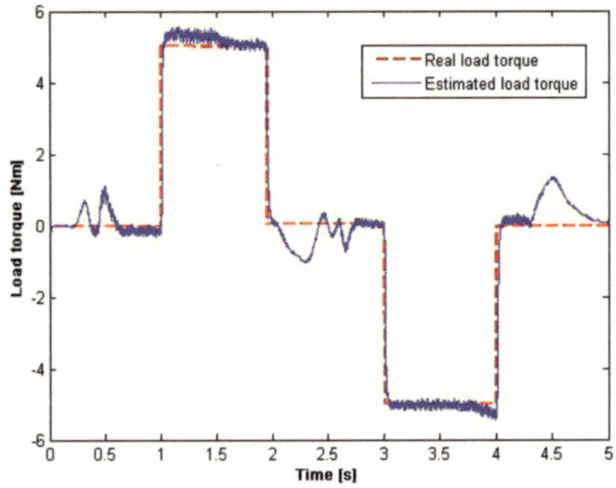

Fig. 5. Reference and estimated load torque

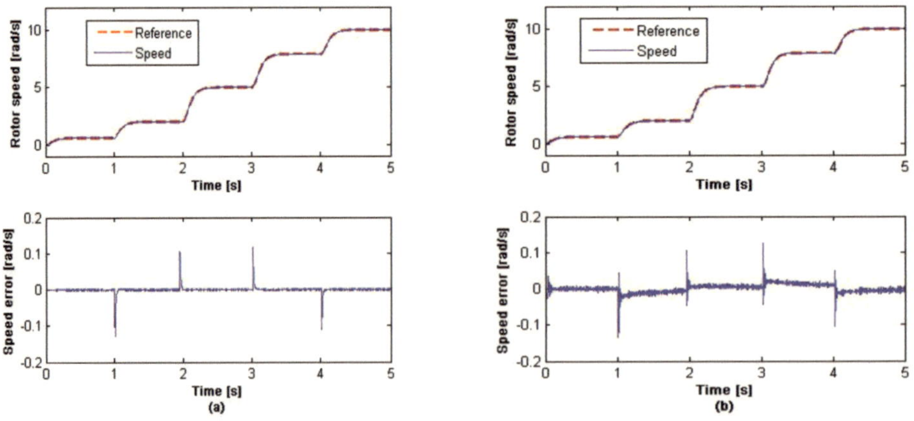

Fig. 6. Low speed tracking performances - (a) proposed NMPC PID Controller, and (b) Field Oriented Controller (FOC).

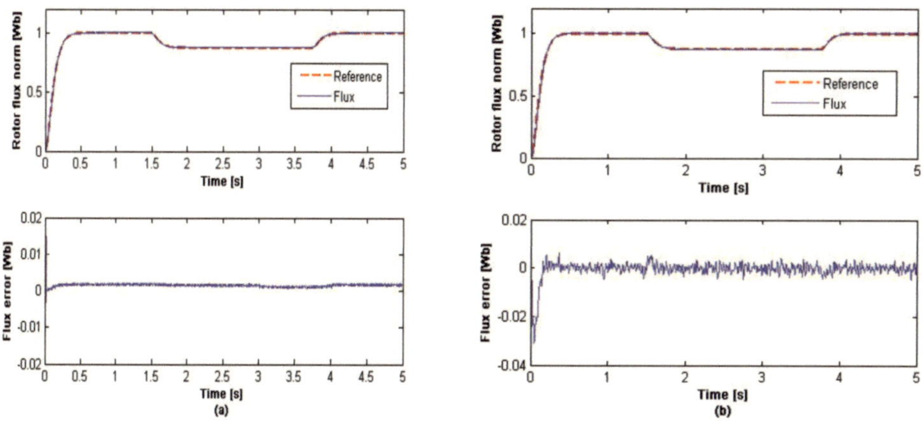

Fig. 7. Flux norm tracking performances for low speed operation - (a) proposed NMPC PID Controller, and (b) Field Oriented Controller (FOC).

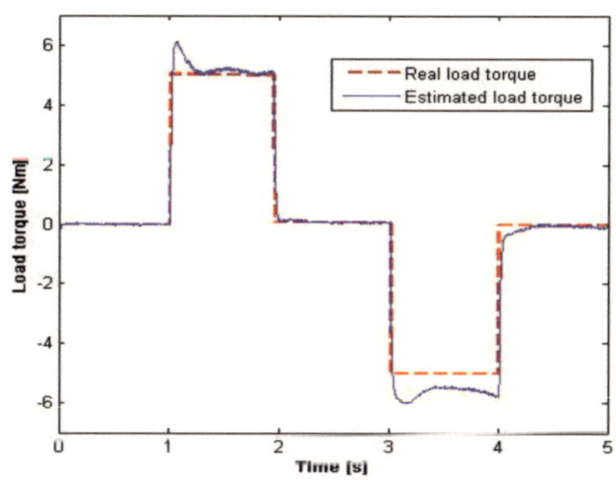

Fig. 8. Reference and estimated load torque

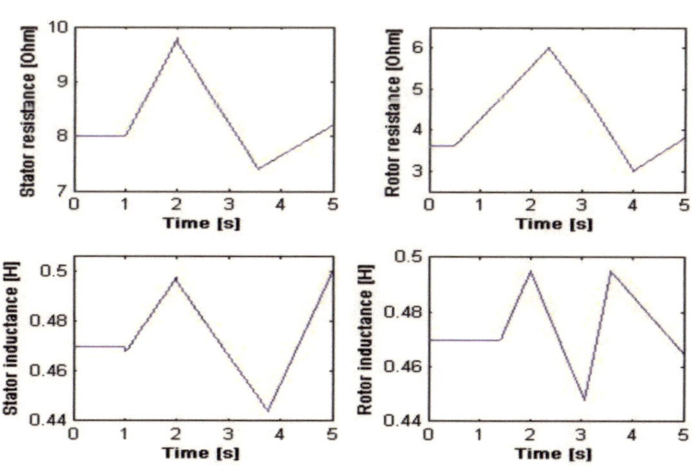

Fig. 9. Variation of machine parameters

Figure 10 gives the tracking responses for speed and flux norm in case of reversal speed. It can be seen that the speed and rotor flux are slightly influenced by the variations. However, the disturbance observation, in figure 11, is deteriorated by the variations. Although a deterioration of perturbation estimation is observed, the tracking of the mismatched model is achieved successfully, and the load torque variations are well rejected in speed response, which is the target application of the drive. Figure 12 gives the tracking responses for speed and flux norm in case of low speed. The speed and rotor flux responses are not affected by the parameters variations. The disturbance observation, shown in figure 13, is less affected than in first case. Although the load torque estimation is sensitive to the speed error, its rejection in speed response is achieved accurately.

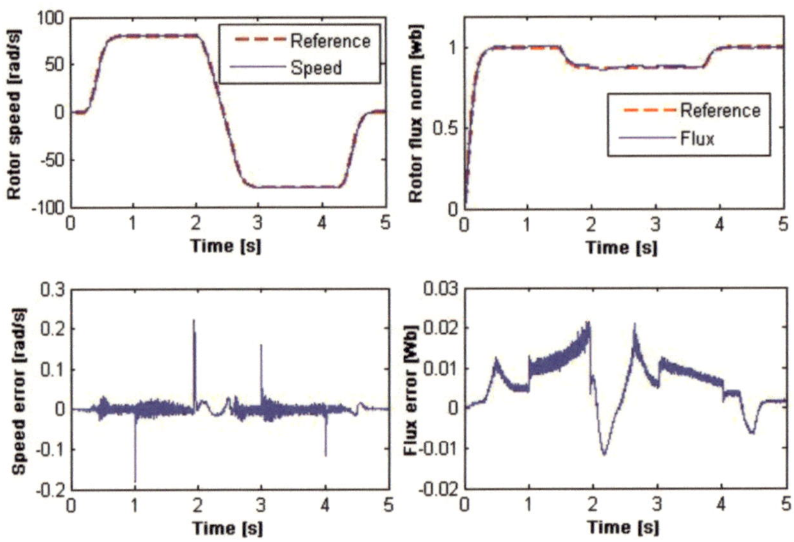

Fig. 10. Speed and flux norm tracking performances under motor parameters variation.

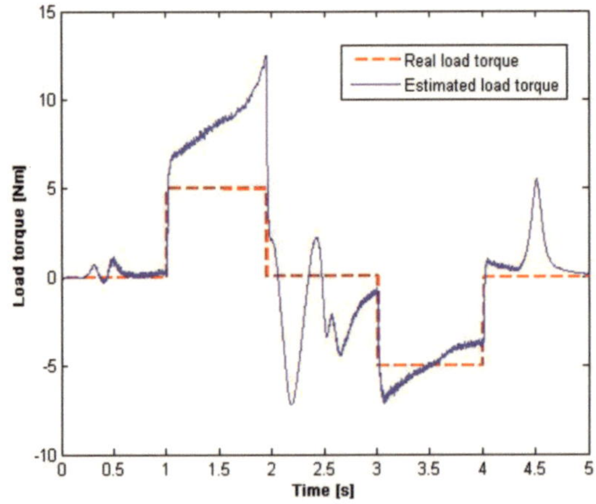

Fig. 11. Reference and estimated load torque under motor parameters variation.

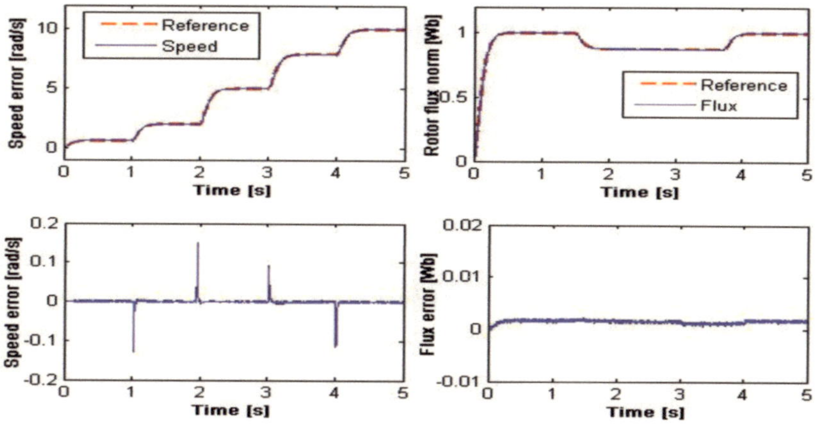

Fig. 12. Speed and flux norm tracking performances under motor parameters variation.

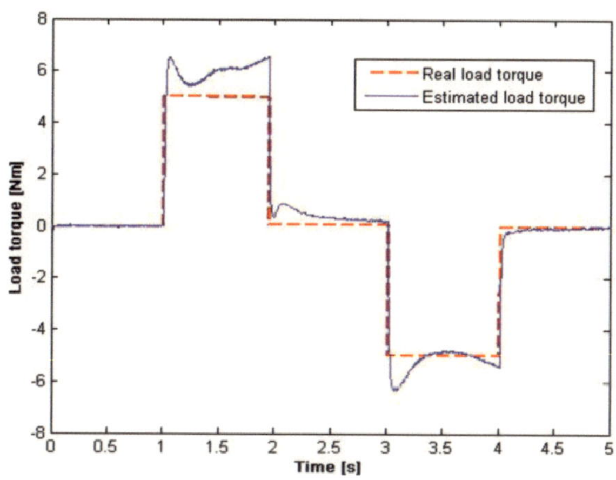

Fig. 13. Reference and estimated load torque under motor parameters variation.

It can be seen that the disturbance observation is influenced by transitions in speed response. Furthermore, the use of the state observer may influence on the system response. Therefore, a more powerful state observer can improve the controlled system performance.

An improvement can be achieved by introduction of an on-line parameters identification, which leads to the adaptive techniques (Marino et al., 1998; Van Raumer, 1994), which is beyond the scope of this chapter.

7. Conclusion

An application of nonlinear PID model predictive control algorithm to induction motor drive is presented in this chapter. First, the nonlinear model predictive control law has been carried out from the nonlinear state model of the machine by minimizing a cost function. Even though the control weighting term is not included in the cost function, the tracking performance is achieved accurately. The computation of the model predictive control law is easy and does not need an online optimization. It has been shown that the stability of the closed loop system under this controller is guaranteed. Then, the load torque is considered as an unknown disturbance variable in the state model of the machine, and it is estimated by an observer. This observer, derived from the nonlinear model predictive control law, is simplified to a PID speed controller. The integration of the load torque observer in the model predictive control law allows enhancing the performance of the motor drive under machine parameter variations and unknown disturbance. The combination between the NMPC and disturbance observer forms the NMPC PID controller. In this application, it has been noticed that the tuning of the NMPC PID controller parameters is easier compared with the standard FOC method.

A state observer is integrated in the control scheme. The global stability of the whole system is theoretically proved using the Lyapunov technique. Therefore, the coupling between the nonlinear model predictive controller and the state observer guarantees the global stability.

The obtained results show the effectiveness of the proposed control strategy regarding trajectory tracking, sensitivity to the induction motor parameters variations and disturbance rejection.

8. Appendices

8.1 Lie derivatives of the process outputs

The following notation is used for the Lie derivative of state function $h_j(\mathbf{x})$ along a vector field $\mathbf{f(x)}$.

$$L_{\mathbf{f}}h_j = \frac{\partial h_j}{\partial \mathbf{x}}\mathbf{f}(\mathbf{x}) = \sum_{i=1}^{n}\frac{\partial h_j}{\partial x_i}f_i(\mathbf{x}) \tag{A1}$$

Iteratively, we have

$$L_{\mathbf{f}}^k h_j = L_{\mathbf{f}}(L_{\mathbf{f}}^{(k-1)}h_j)\,;\, L_{\mathbf{g}}L_{\mathbf{f}}h_j = \frac{\partial L_{\mathbf{f}}h_j}{\partial \mathbf{x}}\mathbf{g}(\mathbf{x}) \tag{A2}$$

$$L_{\mathbf{f}}h_1(\mathbf{x}) = \frac{pL_m}{JL_r}\left(\phi_{r\alpha}i_{s\beta} - \phi_{r\beta}i_{s\alpha}\right) - \frac{f_r}{J}\omega - \frac{1}{J}T_L \tag{A3}$$

$$L_{\mathbf{f}}^2h_1(\mathbf{x}) = \frac{pL_m}{JL_r}\left(\gamma + \frac{1}{T_r} + \frac{f_r}{J}\right)\left(\phi_{r\beta}i_{s\alpha} - \phi_{r\beta}i_{s\alpha}\right) - \frac{p^2L_mK}{JL_r}\left(\phi_{r\alpha}^2 + \phi_{r\beta}^2\right) - \frac{p^2L_m}{JL_r}\omega\left(\phi_{r\alpha}i_{s\alpha} + \phi_{r\beta}i_{s\beta}\right) + \frac{f_r^2}{J^2}\omega + \frac{f_r}{J^2}T_L \tag{A4}$$

$$L_{g_{11}}L_{\mathbf{f}}h_1(\mathbf{x}) = -\frac{pL_m}{J\sigma L_sL_r}\phi_{r\beta} \tag{A5}$$

$$L_{g_{12}}L_{\mathbf{f}}h_1(\mathbf{x}) = \frac{pL_m}{J\sigma L_sL_r}\phi_{r\alpha} \tag{A6}$$

$$L_{\mathbf{f}}h_2(\mathbf{x}) = \frac{2L_m}{T_r}\left(\phi_{r\alpha}i_{s\alpha} + \phi_{r\beta}i_{s\beta}\right) - \frac{2}{T_r}\left(\phi_{r\alpha}^2 + \phi_{r\beta}^2\right) \tag{A7}$$

$$L_{\mathbf{f}}^2h_2(\mathbf{x}) = -\frac{2L_m}{T_r}\left(\gamma + \frac{3}{T_r}\right)\left(\phi_{r\alpha}i_{s\alpha} + \phi_{r\beta}i_{s\beta}\right) - \frac{2pL_m}{T_r}\omega\left(\phi_{r\beta}i_{s\alpha} - \phi_{r\beta}i_{s\alpha}\right) + \frac{4 + 2L_mK}{T_r^2}\left(\phi_{r\alpha}^2 + \phi_{r\beta}^2\right) + \frac{2L_m^2}{T_r^2}\left(i_{r\alpha}^2 + i_{r\beta}^2\right) \tag{A8}$$

$$L_{g_{11}}L_{\mathbf{f}}h_2(\mathbf{x}) = \frac{2L_m}{\sigma L_s T_r}\phi_{r\alpha} \tag{A9}$$

$$L_{g_{12}}L_{\mathbf{f}}h_2(\mathbf{x}) = \frac{2L_m}{\sigma L_s T_r}\phi_{r\beta} \tag{A10}$$

$$L_{g_{21}}h_1(\mathbf{x}) = -\frac{1}{J} \tag{A11}$$

$$L_{g_{21}}L_{\mathbf{f}}h_1(\mathbf{x}) = \frac{f_r}{J^2} \tag{A12}$$

8.2 Simplification of Lie derivatives according l(x)

Using the Lie notations (A1, A2) and output differentiations, in (4) and (6), with **l(x)**, defined by (22), we have

$$L_{g_{11}}L_fh_1(x) = \frac{\partial L_fh_1(x)}{\partial x}g_{11}(x) = \frac{1}{p_0}l(x)g_{11}(x) \tag{B1}$$

$$L_{g_{12}}L_fh_1(x) = \frac{\partial L_fh_1(x)}{\partial x}g_{12}(x) = \frac{1}{p_0}l(x)g_{12}(x) \tag{B2}$$

$$L_{g_{21}}L_fh_1(x) = \frac{\partial L_fh_1(x)}{\partial x}g_{21}(x) = \frac{1}{p_0}l(x)g_{21}(x) - K_1L_{g_{21}}h_1(x) \tag{B3}$$

$$L_f^2h_1(x) = \frac{\partial L_fh_1(x)}{\partial x}f(x) = \frac{1}{p_0}l(x)f(x) - K_1L_fh_1(x) \tag{B4}$$

$$\begin{aligned} l(x)\dot{x} &= p_0\left(\frac{\partial L_fh_1(x)}{\partial x}\frac{\partial x}{\partial t} + K_1\frac{\partial h_1}{\partial x}\frac{\partial x}{\partial t}\right) \\ &= p_0\left(\ddot{\omega}(t) + K_1\dot{\omega}(t)\right) \end{aligned} \tag{B5}$$

$$\mathbf{l}(\mathbf{x})g_2(x) = p_0\left(\frac{\partial L_{\mathbf{f}}h_1(\mathbf{x})}{\partial \mathbf{x}}g_{21} + K_1\frac{\partial h_1}{\partial x}g_{21}\right) = p_0\left(L_{g_{21}}L_{\mathbf{f}}h_1(\mathbf{x}) + K_1L_{g_{21}}h_1(\mathbf{x})\right) = p_0\left(\frac{f_r}{J^2} - K_1\frac{1}{J}\right) = c \tag{B6}$$

8.3 Lie derivatives of the auxiliary outputs

$$\hat{h}_{11} = \frac{pL_m}{JL_r}(\hat{\phi}_{r\alpha}\hat{i}_{s\beta} - \hat{\phi}_{r\beta}\hat{i}_{s\alpha}) \tag{C1}$$

$$\begin{aligned} L_{\hat{\mathbf{f}}}\hat{h}_{11} = \frac{pL_m}{JL_r}[(\gamma + \frac{1}{T_r} + k_1)(\hat{i}_{s\alpha}\hat{\phi}_{r\beta} - \hat{i}_{s\beta}\hat{\phi}_{r\alpha}) - p\hat{\omega}(\hat{i}_{s\alpha}\hat{\phi}_{r\alpha} + \hat{i}_{s\beta}\hat{\phi}_{r\beta}) - pK\hat{\omega}(\hat{\phi}_{r\alpha}^2 + \hat{\phi}_{r\beta}^2) \\ -k_1(i_{s\alpha}\hat{\phi}_{r\beta} - i_{s\beta}\hat{\phi}_{r\alpha}) + \frac{k_2}{T_r}(i_{s\alpha}\hat{i}_{s\beta} - i_{s\beta}\hat{i}_{s\alpha}) - pk_2\hat{\omega}(i_{s\alpha}\hat{i}_{s\alpha} - i_{s\beta}\hat{i}_{s\beta}) + pk_2\hat{\omega}(\hat{i}_{s\alpha}^2 + \hat{i}_{s\beta}^2) - \hat{\phi}_{r\beta}f_{ia} + \hat{\phi}_{r\alpha}f_{ib}] \end{aligned} \tag{C2}$$

$$\hat{h}_{21} = \frac{2L_m}{T_r}\left(\hat{\phi}_{r\alpha}\hat{i}_{s\alpha} + \hat{\phi}_{r\beta}\hat{i}_{s\beta}\right) \tag{C3}$$

$$\begin{aligned} L_{\hat{\mathbf{f}}}\hat{h}_{21} = \frac{2L_m}{T_r}[(\frac{L_m}{T_r} - \frac{k_2}{T_r})(\hat{i}_{s\alpha}^2 + \hat{i}_{s\beta}^2) - (\gamma + \frac{1}{T_r} + k_1)(\hat{i}_{s\alpha}\hat{\phi}_{r\beta} + \hat{i}_{s\beta}\hat{\phi}_{r\alpha}) - p\hat{\omega}(\hat{i}_{s\alpha}\hat{\phi}_{r\alpha} - \hat{i}_{s\beta}\hat{\phi}_{r\beta}) + \frac{k_2}{T_r}(i_{s\alpha}\hat{i}_{s\alpha} + i_{s\beta}\hat{i}_{s\beta}) \\ + pK\hat{\omega}(i_{s\alpha}\hat{i}_{s\beta} - i_{s\beta}\hat{i}_{s\alpha}) + \frac{K}{T_r}(\hat{\phi}_{r\alpha}^2 + \hat{\phi}_{r\beta}^2) + k_1(i_{s\alpha}\hat{\phi}_{r\alpha} + i_{s\beta}\hat{\phi}_{r\beta}) + \hat{\phi}_{r\alpha}f_{ia} + \hat{\phi}_{r\beta}f_{ib}] \end{aligned} \tag{C4}$$

8.4 Induction machine characteristics

The plant under control is a small induction motor 1.1 kW, with the following parameters

ω_{nom} = 73.3 rad/s, $\phi_{r\alpha\beta}$ = 1.14 Wb, T_{nom} = 7 Nm, R_s = 8.0 Ω, R_r = 3.6 Ω, L_s = 0.47 H, L_r = 0.47 H, L_m = 0.44 H, p = 2, f_r = 0.04 Nms, J = 0.06 kgm^2

9. References

Barut, M.; Bogosyan, S.; Gokasan, M. (2005). Speed sensorless direct torque control of IM's with rotor resistance estimation. *Energy Conversion and Management*, 46, pp. 335-349.

Bemporad, A.; Morari, M.; Dua, V.; Pistikopoulous, E.N. (2002). The explicit linear quadratic regulator for constrained systems *Automatica*, 38 (1), pp. 3-20.

Blaschke, F. (1972). The principle of field orientation as applied to the new transvector closed loop system for rotating field machines. *Siemens Rev.*, 39 (5), pp. 217-220.

Bordons, C.; Camacho, E.F. (1998). A generalized predictive controller for a wide class of industrial processes. *IEEE Transactions on Control Systems Technology*, 6 (3), pp. 372-387.

Camacho, E.F.; Bordons, C. (2004). Model Predictive Control, 2nd edition, Springer.

Chen, F.; Dunnigan, M.W. (2003). A new non-linear sliding-mode for an induction motor machine incorporating a sliding-mode flux observer. *International Journal of Robust and Nonlinear Control*, 14, pp. 463-486.

Chen, W.H.; Balance, D.J.; Gawthrop P.J. (2003). Optimal control of nonlinear systems: a predictive control approach. *Automatica*, 39 (4), pp. 633-641.

Chen, W.H.; Balance, D.J.; Gawthrop, P.J.; Gribble, J.J.; O'Reilly J. (1999). Nonlinear PID predictive controller. *IEE Proceedings Control Theory Application*; 146 (6), pp. 603-611.

Chenafa, M.; Mansouri, A.; Bouhenna, A.; Etien, E.; Belaidi, A. & Denai, M.A. (2005). Global stability of linearizing control with a new robust nonlinear observer of the induction motor. *International Journal of Applied Mathematics and Computer Sciences, 15* (2), pp. 235-243.

Chiasson, J. (1996). Nonlinear controllers for an induction motor. *Control Engineering Practice*, 4 (7), pp. 977-990.

Correa, P.; Pacas, M.; Rodriguez, J. (2007). Predictive Torque Control for Inverter Fed Induction Machines. *IEEE Transactions on Industrial Electronics*, 45 (2), pp. 1073-1079.

Du, T.; Brdys, M.A. (1993). Shaft speed, load torque and rotor flux estimation of induction motor drive using an extended Luenberger observer. 6th IEEE International Conference on Electrical Machines and Drives, pp. 179-184.

Feng, W.; O'Reilly, J.; Balance D.J. (2002). MIMO nonlinear PID predictive controller. *IEE Proceedings Control Theory Application*, 149 (3), pp. 203-208.

Garcia, C. E.; Prett, D.M., Morari, M. (1989). Model predictive control: theory and practice- a survey. *Automatica*, 3, pp. 335-348.

Hedjar, R.; Toumi, R.; Boucher, P.; Dumur D. (2003). Two cascaded nonlinear predictive control of induction motor. Proceedings of the IEEE Conference on Control Application, Istanbul, Turkey; 1, pp. 458-463.

Hedjar, R.; Toumi, R.; Boucher, P.; Dumur, D. (2000). Cascaded nonlinear predictive control of induction motor. Proceedings of the IEEE Conference on Control Applications, Anchorage, Alaska, USA, pp. 698-703.

Hong, K.; Nam, K. (1998). A load torque compensation scheme under the speed measurement delay. *IEEE Transactions on Industrial Electronics*, 45 (2), 283-290.

Jansen, L.P.; Lorenz, D.W. Novotny (1994). Observer-based direct field orientation: analysis and comparison of alternatives methods. *IEEE Transactions on Industry Applications*, 30 (4), pp. 945-953.

Leonhard, W. (2001). Control of Electrical Drives. 3rd Edition, Spinger-Verlag: Berlin.

Maaziz, M.K.; Boucher, P.; Dumur, D. (2000). A new control strategy for induction motor based on non-linear predictive control and feedback linearization. *International Journal of Adaptive Control and Signal Processing*, 14, pp. 313-329.

Marino, R.; Peresada, S.; Tomei, P. (1998). Adaptive output feedback control of current-feed induction motors with uncertain rotor resistance and load torque. *Automatica*, 34 (5), pp. 617-624.

Marino, R.; Peresada, S.; Valigi, P. (1993). Adaptive input-output linearizing control of induction motors. *IEEE Transactions on Automatic Control*, 38 (2), pp. 208-221.

Marino, R.; Tomei, P.; Verrelli, C.M. (2002). An adaptive controller for speed-sensorless current-feed induction motors with unkown load torque. Proceedings of the 7th International Conference on control, Automation, Robotics and Vision, Singapore, pp. 1658-1663.

Merabet, A.; Ouhrouche, M.; Bui, R.T. (2006). Nonlinear predictive control with disturbance observer for induction motor drive. Proceedings of IEEE International Symposium on Industrial Electronics, Montreal, Canada.

Nemec, M.; Nedeljkovic D.; Ambrozic; V. (2007). Predictive Torque Control of Induction Machines Using Immediate Flux Control. *IEEE Transactions on Industrial Electronics*, 54 (4), pp. 2009-2017.

Ping L. (1996). Optimal predictive control of continuous nonlinear systems. *International Journal of Control*; 63 (1), pp. 633-649.

Richalet, J. (1993). Industrial applications of model based predictive control. *Automatica*, 29 (5), pp. 1251-1274.

Siller-Alcalá, I.I. (2001). Generalized predictive control for nonlinear systems with unstable zero dynamics. *Journal of the Mexican Society of Instrumentation and Development*, 5 (3), pp. 146-151.

Van Raumer, T. (1994). Nonlinear adaptive control of induction machine. PhD thesis (in French), LAG, Grenoble, France.

Development of Real-Time Hardware in the Loop Based MPC for Small-Scale Helicopter

Zahari Taha[2], Abdelhakim Deboucha[1],
Azeddein Kinsheel[1] and Raja Ariffin Bin Raja Ghazilla[1]
[1]*Centre for Product Design and Manufacturing Department of Engineering Design, Manufacture Faculty of Engineering- University of Malaya, Kuala Lumpur,*
[2]*Department of Manufacturing Engineering, University Malaysia Pahang, Gambang, Pahang, Malaysia*

1. Introduction

In recent years, unmanned aerial vehicles (UAVs) have been shown a rapid development equipped with intelligent flight control devices. Many advantages could be offered by UAVs, due to their widely applications (Garcia and Valavanis, 2009).

Flight control is the principle unit for UAVs to perform a full autonomous mission without or with less interference of a human pilot. Numerous types of control have been developed for small-scale helicopters including classical, intelligent and vision controls.

The most conventional and common control methods that have been used by many researchers are the SISO controls, i.e., PI or PID control because their requirements are not highly dependent on the accuracy of the plant model. Two control approaches are proposed by Kim and Shim (2003), a multi-loop PID controller and a nonlinear model predictive control for the purpose of trajectory tracking. This strategy shows satisfactory results when applied to the Yamaha R-50. However, if large perturbations need to be compensated, or significant tracking abilities are required, this strategy may not be adequate.

Wenbing, et al (2007) presented a multiple-input-multiple-output (MIMO) neural network controller which has a structure of two inputs and two outputs to control a small-scale helicopter. The neural network controller is used with a simple adaptive PID configuration. The PID gains k_i, k_p and k_d are tuned online via the training of the proposed neural networks during the flight.

Srikanth, James and Gaurav (2003) combined vision with low-level control to perform the landing operation of the helicopter. The vision control navigates the commands to a low-level controller to achieve a robust landing. In their experiments, state initializations were set in the hover condition arbitrarily. The idea was to find the helipad, then align with it, and land on it. The low-level (roll, pitch lateral and heading) controls were implemented with a proportional controller. The altitude behaviour was implemented with a PI controller. To make altitude control easier, the PI controller was split into three: sub-hover control, velocity and sonar control.

In Montgomery et al (1995), the control system proposed in the USC architecture is implemented as PD control loops with gains tuned by trial and error. In hovering conditions, the system is assumed linear (or linearized), thus multivariable linear control techniques such as Linear Quadratic Regulator (LQR) and H_∞ can be applied. Edgar, Hector and Carlos (2007) propose a flight control structure by combining PID, fuzzy and regulation control, using a nonlinear MIMO model for an X-Cell mini-helicopter platform.

Recently, intelligent control methods have become popular and an alternative to conventional methods. Intelligent control methods can act efficiently with nonlinear and unstable systems. In general, these methods can be categorized into three main techniques: fuzzy control, neural networks approach and genetic algorithm. Furthermore, these techniques can be combined with each other or with conventional methods to become hybrid techniques.

The genetic algorithm based on floating point representation has been modified to tune the longitudinal controller gains parameters of a two-bladed XCELL helicopter platform by Mario, G. P. (1997). First principle modelling is used to model the longitudinal behaviour of the platform. The author applied and compared the proposed design in both time and frequency domains. This algorithm shows faster convergence of the system with less computational time. Kadmiry and Driankov (2003) propose a combination of a Fuzzy Gain Scheduler (FGS) and a linguistics (Mamdani-type) controller. The authors used the FGS to control the attitude stability of the helicopter, whereas the linguistics controller was used to generate the inputs to the fuzzy controller for the given attitudes (z, roll, pitch, and yaw). The proposed controller scheme contains two loops; outer lop and inner loop. The inner loop represents the attitude controller and the outer loop deals with the translational rate variables. The controller was obtained and simulated based on a real nonlinear dynamic model of the platform.

This paper addresses the control problem of the HIROBO model platform which is being developed by University of Malaya team. The details of the system hardware and data collection are presented by (Zahari, et al, 2008) and (Taha. Z, T et al, 2010), respectively. The black box Nonlinear Autoregressive Model (NARX) modelling and identification of the platform is presented by (Deboucha,. A et al, 2010). The use of this model was preferred because of its ability to handle instability and nonlinearity of complex nonlinear dynamic and unstable systems such as the helicopter. The author estimated the NARX based on collected flight data test (Taha. Z, T et al, 2010). In this paper, the obtained model by Deboucha. A, et al (2010) is used as plant model to be controlled. Due to the complementary between Model Predictive Control (MPC) and NARX, the MPC algorithm is applied to control the stability of the helicopter. MPC algorithm differs from other control strategies in: firstly, its multi-variables feature and secondly the possibility of using constraints. Therefore, reasonable results are anticipated. To prove the capabilities of the latter control, it has been simulated as model in the loop using SIMULINK. Furthermore, an xpctraget rapid prototype is developed to implement and test the controller to play the role of hardware in the loop test (HIL).

2. Model description

In this section, a brief description of the NARX black box model is presented. As reported previously, the identified orientation model of a Hirobo scale helicopter is obtained by (Deboucha, .A, et al, 2010). A standard NARX discrete time nonlinear multivariable model system with m outputs and r inputs is a general parametric form for modelling Black-box nonlinear systems with one step ahead prediction, which can be described by the following formula (Zhang & Ljung, 1999).

$$\hat{y}_m(k) = N\,[\,y_m(k-1), \ldots, y_m(k-n_a), u_r(k-n_k), \ldots, u_r(k-n_k-n_b+1)] + e_m(k) \qquad (1)$$

where, $y_m = [\emptyset\ \theta\ \varphi]^{\mathrm{T}}$ are the orientation behaviour of the helicopter and $u_r = [u_{elev}, u_{aile}, u_{colle}, u_{ped}]^T$, are the swash-plate control input vector. n_a, n_b are the matrices of the past outputs and inputs involved in the system, respectively, n_k is a matrix of the input delays from each input to each output. *N(.)* represents unknown nonlinear function, which in this case is computed by the neural network technique for estimating the nonlinearity of the system. Since the input vector to the system is from the swash-plate, the dynamics of the actuators i.e. servo positions are not included. The dynamics from actuators position to the swash-plate control inputs is assumed linear. This mapping model is presented by (Deboucha, .A, et al, 2010).

The model which is presented in LTI state space form is linearized about a specified input vector and treated in terms of stability to ensure the performance of the model.

The linearized LTI state space model of the orientation dynamics of the helicopter platform is given by:

$$\begin{cases} x_m(k+1) = A_m x_m(k) + B_m u_r(k) \\ y_m(k) = C_m x_m(k) + D_m u_r(k) \end{cases} \qquad (2)$$

where, $u_r(k)$ is the manipulated variable (control inputs to the helicopter), $y_m(k)$is the process output which are the Euler angles and $x_m(k)$ is the state vector.

The state space matrices have obtained by (Deboucha., A ,2011) as,

$$A_m = \begin{bmatrix} 0.9994 & 0.02606 & -0.8908 & -0.1565 & -1.019 & -0.0052 & -0.7776 & -1.774 & 0.5769 \\ 0.002018 & 1 & 0.4515 & -0.1531 & 0.1697 & -0.2016 & 0.1594 & 0.2529 & -0.3566 \\ 0.006199 & 0.01822 & -0.03431 & 0.8172 & -0.2367 & 0.03092 & -0.1036 & -0.165 & 0.05891 \\ 0.005672 & 0.01667 & 0.115 & 0.1182 & -0.2007 & 0.0178 & 0.03018 & -0.1787 & 0.1455 \\ -0.01055 & -0.03101 & -0.3068 & 0.1792 & 0.3507 & -0.03528 & -0.4128 & -0.6006 & -0.00849 \\ -0.001958 & -0.005753 & -0.05692 & 0.03324 & 0.08312 & 0.006115 & 0.3217 & -0.2741 & -0.9223 \\ 0.01661 & 0.04881 & 0.483 & -0.282 & -0.5697 & 0.04318 & 0.2607 & -0.4691 & 0.4382 \\ -0.002215 & -0.006509 & -0.0644 & 0.03761 & 0.07598 & -0.005758 & -0.03476 & 0.06961 & -0.0563 \\ 0.007343 & 0.02158 & 0.2135 & -0.1247 & -0.2519 & 0.01909 & 0.1152 & -0.2308 & 0.1866 \end{bmatrix}$$

$$B_m = \begin{bmatrix} -0.04578 & -0.07926 & -0.02391 & 0.1732 \\ 0.001391 & 0.002409 & 0.0007266 & -0.005263 \\ 0.266 & 0.06843 & 0.02064 & -0.1495 \\ -0.9375 & 0.0626 & 0.01889 & -0.1368 \\ -0.05145 & -0.1165 & -0.03514 & 0.2998 \\ -0.03485 & -0.02161 & -0.00652 & 0.07897 \\ 0.1059 & 0.1834 & 0.05529 & 0.5978 \\ -0.01412 & -0.02441 & 0.9926 & 0.05346 \\ 0.04682 & -0.9189 & 0.02449 & -0.1771 \end{bmatrix}$$

$$C_m = \begin{bmatrix} -2.473 & -7.36 & 3.98 & 0.2138 & -7.225 & -8.439 & 1.357 & 0.9117 & 4.069 \\ 1.657 & 0.04449 & -1.599 & -0.04054 & 0.1303 & 0.5443 & -1.073 & -3 & 0.8277 \\ -0.1586 & -6.489 & 1.333 & 3.037 & 5.237 & 0.1375 & -4.339 & 0.3119 & -1.192 \end{bmatrix}$$

$$D_m = \begin{bmatrix} 0 & 0 & 0 & 0 \\ 0 & 0 & 0 & 0 \\ 0 & 0 & 0 & 0 \\ 0 & 0 & 0 & 0 \end{bmatrix}$$

3. Control design

The objective of the MPC in this study is to bring the helicopter to its equilibrium i.e. the hovering condition. The controller is designed in the case where the translation velocities are decaying to zeros and the Euler angles are limited with specific constraints explained in the next section. With these parameter criteria, the helicopter tries to stabilize into a hover state.

To design an MPC control, the above matrices have to be updated following the procedure below addressed in (Liuping, 2008 and Jay, et.al, 1994).

By taking the difference operation in both sides of the formula (2)

$$x_m(k+1) - x_m(k) = A_m\big(x_m(k) - x_m(k-1)\big) + B_m(u_r(k) - u_r(k-1)) \tag{3}$$

Then by defining

$$\begin{cases} \Delta x_m(k+1) = x_m(k+1) - x_m(k) \\ \Delta x_m(k) = x_m(k) - x_m(k-1) \\ \Delta u(k) = u_r(k) - u_r(k-1) \end{cases} \tag{4}$$

the updated states model would be as follows:

$$\Delta x_m(k+1) = A_m \Delta x_m(k) + B_m \Delta u_r(k) \tag{5}$$

From (2) and (5), the relation between the outputs of the system and the state variables could be deduced as

$$\begin{aligned} \Delta y_m(k+1) &= C_m \Delta x_m(k+1) + D_m \Delta u_r(k+1) \\ &= C_m A_m \Delta x_m(k) + C_m B_m \Delta u_r(k) + D_m \Delta u_r(k+1) \end{aligned} \tag{6}$$

It can also defined as

$$\Delta y_m(k+1) = y_m(k+1) - y_m(k) \tag{7}$$

The augmented state space model has a new state defined by

$$x(k) = [\Delta x_m(k)^T \; y_m(k)^T]^T \;, \tag{8}$$

where, the predicted state space model is deduced as

$$\begin{bmatrix} \Delta x_m(k+1) \\ y_m(k+1) \end{bmatrix} = \begin{bmatrix} A_m & o_m \\ C_m A_m & I_{mxm} \end{bmatrix} \begin{bmatrix} \Delta x_m(k) \\ y_m(k) \end{bmatrix} + \begin{bmatrix} B_m \\ C_m B_m \end{bmatrix} \Delta u_r(k) + \begin{bmatrix} o_m \\ D_m \end{bmatrix} \Delta u_r(k+1) \tag{9}$$

Because D_m is a zero matrix, the last term in the above equation can be eliminated, where, o_m is zeros matrix and I_{mxm} is the identity matrix.

To predict the future behaviour of the system$x(k+1)$, the current information of the plant model should be given by$x(k)$. Thus, the future control signals can be expressed by the following:

$$\Delta u_r(k), \Delta u_r(k+1), \ldots, \Delta u_r(k+N_c-1)\,, \tag{10}$$

where N_c is the control horizon dictating the number of parameters used to capture the future control trajectory. With the given information of the model $x(k)$, the future state variables are predicted for N_p number of samples as

$$x(k+1 \mid k), x(k+2 \mid k), \ldots, x(k+p \mid k), \ldots, x(k+Np \mid k)\,, \tag{11}$$

where $x(k + p| k)$ is the predicted state variable at $k + p$ with given current plant $x(k)$.

Based on the predicted state space model with the matrices (A, B, C), where $A = \begin{bmatrix} A_m \\ C_m A_m \end{bmatrix}$, $B = \begin{bmatrix} o_m \\ I_{mxm} \end{bmatrix}$, $C = \begin{bmatrix} B_m \\ C_m B_m \end{bmatrix}$

the forward state variables could be calculated sequentially and finalized for a sample Np as

$$x(k+Np \mid k = A^{Np}\, x(k) + A^{Np-1}\, Bx(k) + A^{Np-2}\, Bx(k+1) + \cdots + A^{Np-N_c}\, Bx(k+N_c-1) \quad (12)$$

Similarly, from the predicted output state variables (8), the predicted outputs are written as follows:

$$y_m(k+Np \mid k) = CA^{Np}x(k) + CA^{Np-1}x(k) + CA^{Np-2}B\Delta u_q(k+1) + \cdots \\ + CA^{Np-N_c}B\Delta u_q(k+N_c-1) \quad (13)$$

All predicted variables are formulated in terms of the information on current state variable $x(k)$ and the future control $\Delta u(k+j)$, where $j = 1,2, \quad N_c - 1$

$$y_m = [y_m(k+1 \mid k)\; y_m(k+2 \mid k)\; y_m(k+3 \mid k), \ldots, y_m(k+Np \mid k)]^T \quad (14)$$

$$\Delta U = [\Delta u_q(k)\; \Delta u_q(k+1)\; \Delta u_q(k+2) \ldots \Delta u_q(k+N_c-1)]^T \quad (15)$$

From the above formulas, the output vector is concluded as follows:

$$Y = Fx(k) + Q\Delta U \quad (16)$$

To sum up, the predictive model of the helicopter's attitude is updated in order to deal with the MPC design. In the next section the optimization algorithm of the MPC is treated based on a given set point (reference model).

3.1 Control optimization

For a given set point for Euler angles $Rs(k)$ at sample time k within a prediction horizon Np, the objective of the MPC system is to bring the predicted output behaviour of the helicopter as close as possible to the set-point signals, where firstly, assuming that the set-point signals remain constant in the optimization window. This objective is then translated into a design to find the 'best' control parameter vector ΔU such that an error function between the set-points and the predicted outputs is minimized.

Assuming the vector set-point data as $Rs = [r_1(k)\; r_2(k)\; r_3(k)]^T$,

where $r_1(k)$ $r_2(k)$ and $r_3(k)$ are the set points of roll angle, pitch angle and yaw angle, respectively. The cost function J which reflects the control objective is defined as

$$J = (Rs - Y)^T(Rs - Y) + \Delta U^{\mathrm{T}}\, \bar{\mathrm{R}}\Delta \mathrm{U} \quad (17)$$

where the first term is linked to the objective of minimizing the error between the predicted output vector and the set-point signals, while the second term concerns the consideration of ΔU when the objective function J is as small as possible. $\bar{R}$ is a diagonal matrix in the form that $\bar{R} = r_w I_{N_c x N_c}$ ($r_w \geq 0$) where r_w is used as tuning parameter for the desired closed loop performance. The goal would be solely to make the error $(Rs - Y)^T(Rs - Y)$ as small as possible. In the case of large r_w the cost function is interpreted as the situation where would carefully consider how large the ΔU might be and cautiously reduce the error

$$(Rs - Y)^T(Rs - Y)$$

To find the optimal ΔU that will minimize the cost function J
we have $y = Fx(k) + Q\Delta U$,

$$\text{where} \qquad F = \begin{bmatrix} CA \\ CA^2 \\ CA^3 \\ \vdots \\ CA^{N_p} \end{bmatrix} Q = \begin{bmatrix} CA & 0 & \dots 0 \\ CAB & CA & \dots 0 \\ CA^2B & CAB & \dots 0 \\ \vdots & & 0 \\ CA^{N_p-1}B & CA^{N_p-2}B & \dots \quad CA^{N_p-N_c}B \end{bmatrix} \tag{18}$$

and, $J = (Rs - Fx(k))^T (Rs - Fx(k)) - 2\Delta U^T Q^T (Rs - Fx(k)) + \Delta U^T (Q^T Q + \bar{R})\Delta U$
From the first derivative of the cost function J

$$\frac{\partial J}{\partial \Delta U} = -2Q^T(Rs - Fx(k) + 2(Q^T Q + \bar{R})\Delta U \tag{19}$$

The required condition is $\frac{\partial J}{\partial \Delta U} = 0$
From which the optimal solution for the control signal is found:

$$\Delta U = (Q^T Q + \bar{R})^{-1} * Q^T(Rs - Fx(k)) \tag{20}$$

where $(Q^T Q + \bar{R})^{-1}$ is the Hassian Matrix in the optimization.

The MPC is designed based on the above updated model and optimized cost function. One of the criteria of MPC, the constraints in both inputs and outputs has been chosen with regard to the behaviour of the vehicle during the flight test. Based on the flight test, the range outputs that guarantee the behaviour of the helicopter in hovering condition were approximately ± 15 degrees while the input ranges were approximately ± 10 degrees.

The second feature that has to be defined in designing an MPC is the selection of the prediction horizon Np and the control horizon Nc. In this work, the chosen Np value is 25 and the chosen Nc is five-control inputs horizon.

Based on these criteria and to satisfy the equation ($\frac{\partial J}{\partial \Delta U} = 0$), the best tuning output weights to stabilize the model were found to be the following matrix:

$$Q^T Q = \begin{bmatrix} 1.1735 & 0 & 0 \\ 0 & 1.1735 & 0 \\ 0 & 0 & 1.1735 \end{bmatrix} \tag{21}$$

While, the best input weights were found to be

$$\bar{R} = \begin{bmatrix} 0.08521 & 0 & 0 & 0 \\ 0 & 0.08521 & 0 & 0 \\ 0 & 0 & 0.08521 & 0 \\ 0 & 0 & 0 & 0.08521 \end{bmatrix} \tag{22}$$

To validate the previously designed controller that stabilizes the helicopter, the designed MPC is implemented into the same hardware described in the previous section. Using xpc-target software in SIMULINK, the model (in Fig 1) have been developed and deployed into the target PC (PC-104). The model contains the IMU sensor software, the MPC, and the corresponding C/T blocks for both capturing and generating PWM signals. The IMU DATA RECEIVE software reads the behaviour of the helicopter (angular position, acceleration...etc) and sends these data to the controller. The MPC generates the required swash-plate angles as well the pedal control by setting the corresponding servo positions. The relationship between the swash plate which has 120^0 layout and the servo position is given by the transformation matrix from as described by (Deboucha et al, 2010):

$$e = \begin{bmatrix} 0.3333 & 0.3333 & 0.3333 \\ 0.5000 & 0 & -0.5000 \\ 0.3333 & -0.6667 & 0.3333 \end{bmatrix}$$

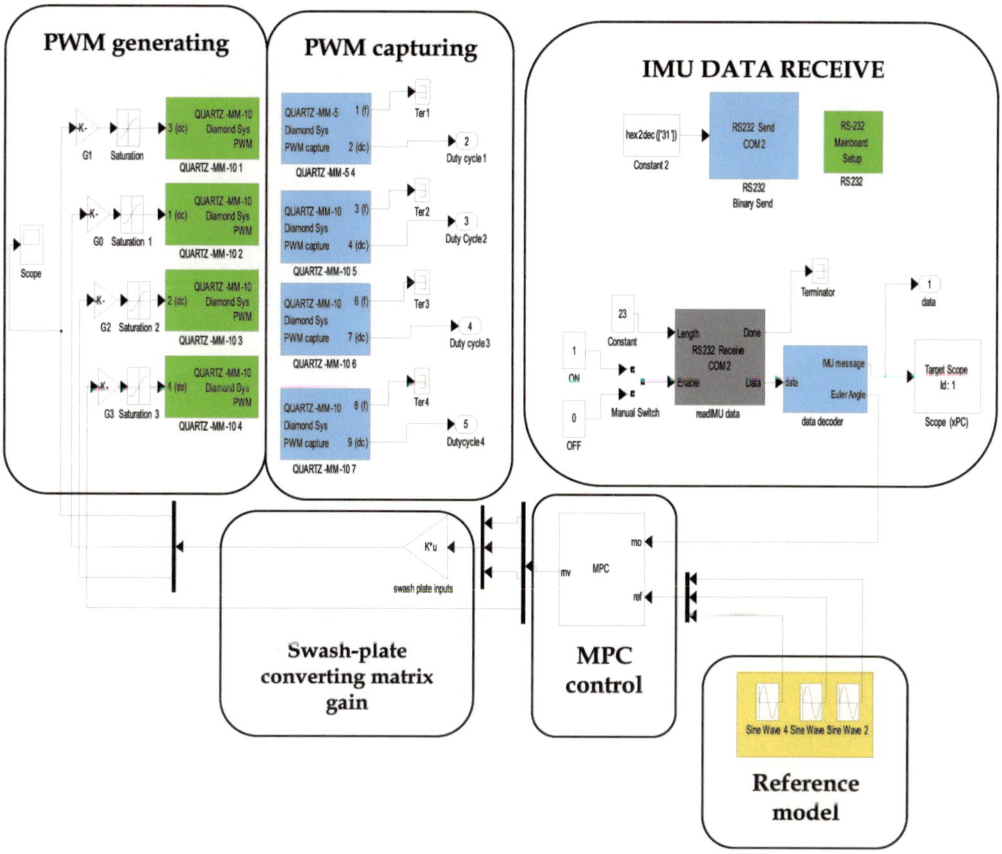

Fig. 1. Simulink Block Diagram model of the MPC implementation

The servos' positions are controlled through a set of PWM signals as described by (Deboucha et al, 2011). The pedal control has no effect on the swash-plate layout. Thus, its PWM signal is determined from the servo position. The corresponding C/T blocks (QUARTZ MM) in the SIMILINK model set the frequency of the PWM signals, which is 50 Hz. To capture the actual PWM signals, the QUARTZ PWM capturing block is used for each generated signal from the MPC. The saturation blocks are used to limit the duty cycle if any over-range of its values.

The main hardware used in this work are 1): A host computer, 2): PC-104, 3): Counter/Timer I/O board, 4): 3DM-GX1 Inertial measurement Unit, 5): two onboard servos (Futaba S3001and Futaba S9254), and 6): Helicopter platform. The sampling time used for the experiment is 0.03s.

4. Experimental setup

A part of this work is to implement the above Simulink model to the target PC-104. Fig.2 presents the overall experimental prototype setup. The IMU sensor was mounted on the

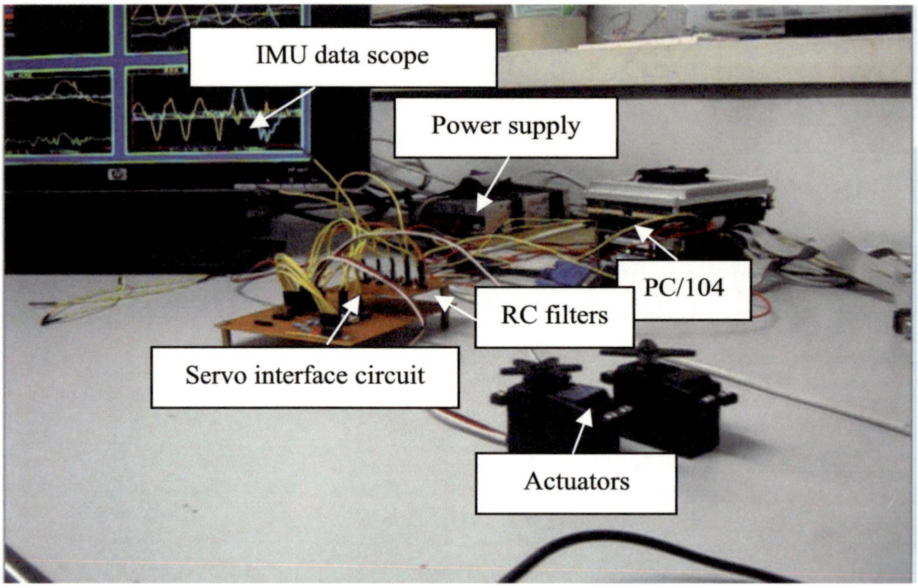

(a): xpctarget prototype configuration

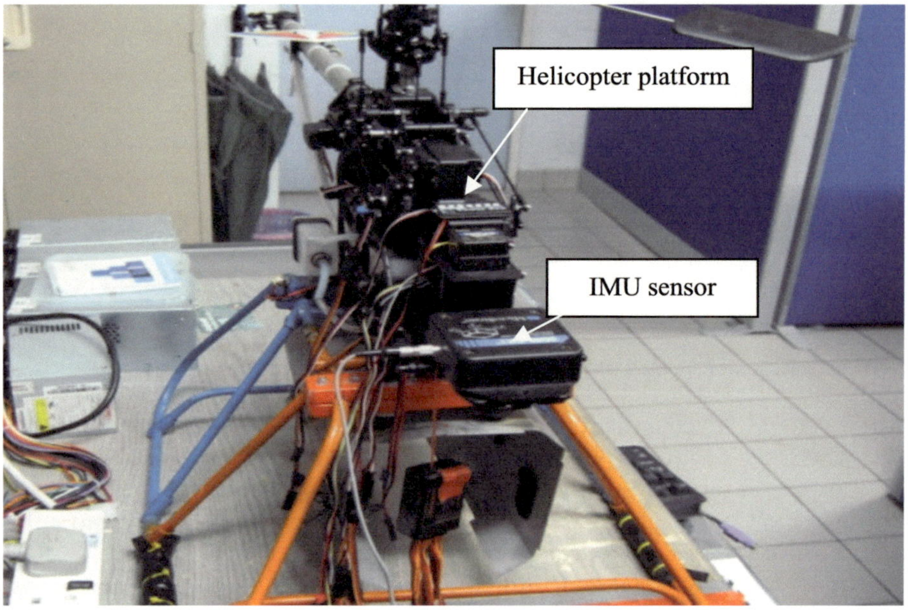

(b): Helicopter platform

Fig. 2. Experimental setup

nose of the helicopter and connected to the target PC-104 through a serial port. The corresponding pins of the I/O counter/timer board were connected to the servo actuators via a servo interface circuit. This circuit is an RC filter used to protect the I/O board form noises produced by the Helicopter components such as the actuators. The PC-104's processor runs the developed system in real time operating system.

5. Simulation results

To test the designed MPC, a simulation of the helicopter performance under different set-points is studied. The step response of the helicopter with the introduction of the disturbance in the roll angle is presented in Fig.3. The amplitudes of the roll, pitch and yaw angles are 12, 10 and 13 degrees, respectively. At 5 seconds, a disturbance with amplitude of 5 degrees is introduced. It can be seen that the controller damps down the amplitude of the angle to 13.3 out of 17 degrees in approximately 5 seconds. The effect of the disturbance on the other states is less and it would appear that there is a small steady state error in the yaw angle after 5 seconds of simulation.

The designed MPC has also been tested to track a square wave with a variety of amplitudes for Euler angles. The performance of the controller is good for all the utilized amplitudes, as illustrated in Figures 4, 5 and 6.

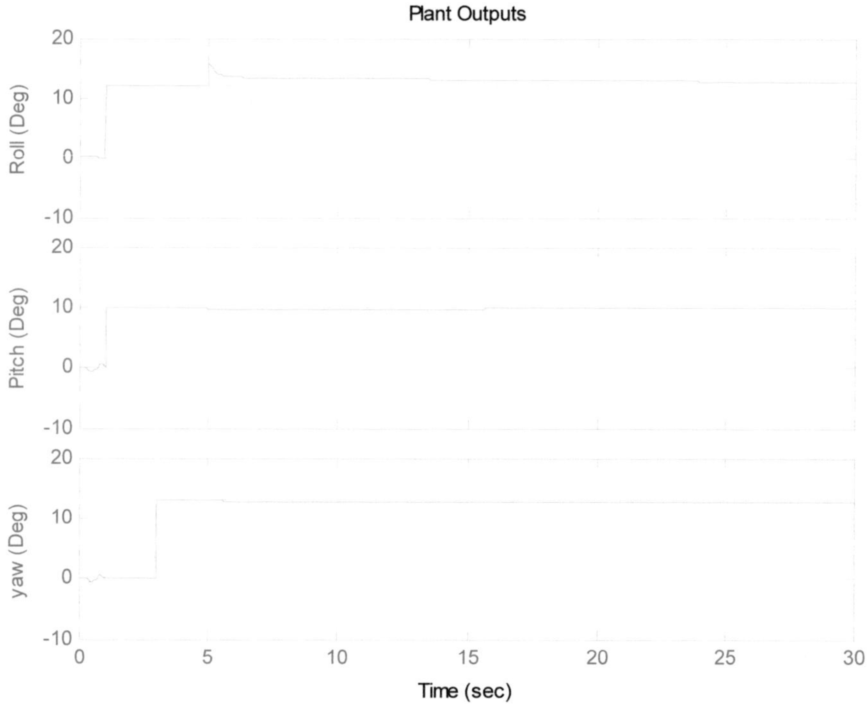

Fig. 3. Step responses

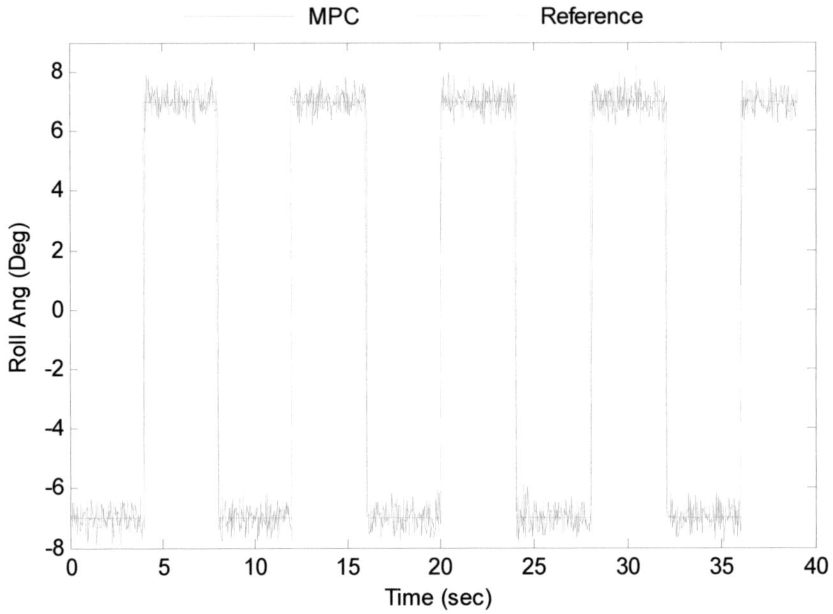

Fig. 4. Square wave roll angle tracking with the MPC controller.

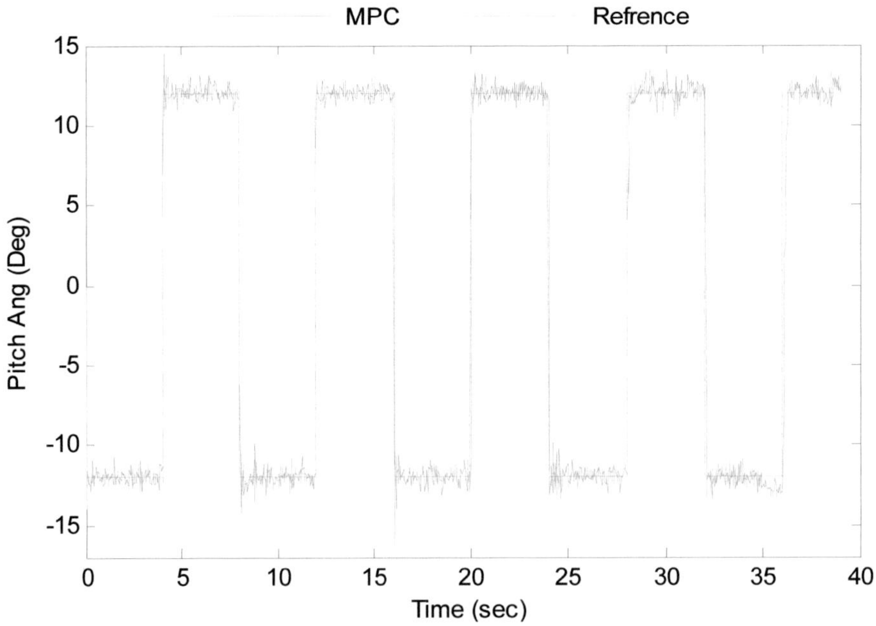

Fig. 5. Square wave pitch angle tracking with the MPC controller

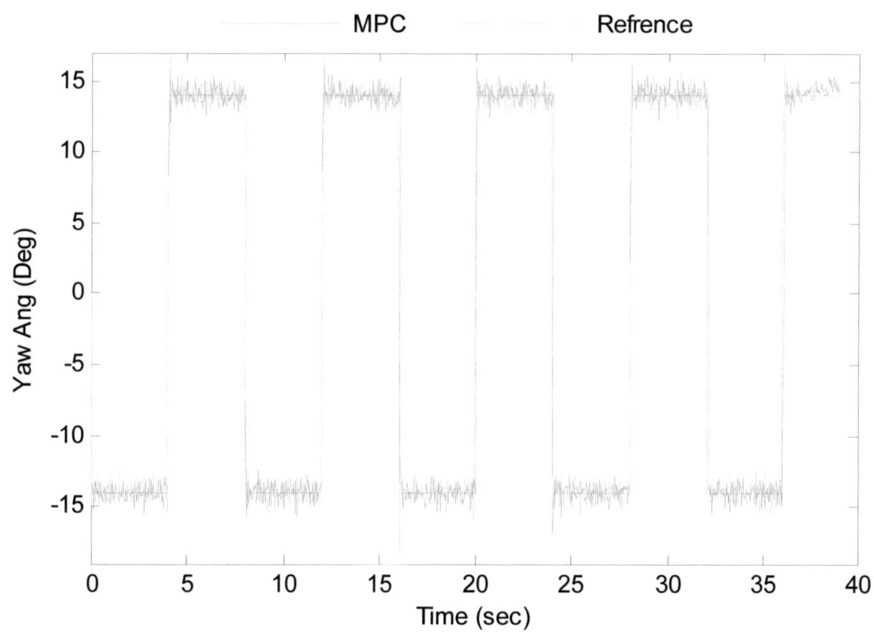

Fig. 6. Square wave yaw angle tracking with the MPC controller.

5.1 Experiment results

This section presents the implementation of developed hardware in the loop system. Two experiments were conducted in this work. The first is conducted where the flight test data were used as reference model and disabling the role of the IMU. Figures 4, 5 and 6 present the generated inputs by the real time MPC to the system to follow the reference model compared with the given inputs system during the flight test. From Fig 1, the collected PWM signals are collected as duty cycle; therefore it has to be transferred to the corresponding angles for each actuator. Instead of activating the IMU software, the feedback to the MPC is the reference model itself. It is noticeable that the generated inputs by the MPC do not follow closely the actual inputs used for modelling task. This is because the MPC is designed based on linearized model of the platform.

As preliminary step to investigate a real autonomous flight, a second experiment is carried out where the IMU software is enabled (fig1) to test its functioning and also to assess how the MPC is sensitive with disturbances. To achieve these criterions, the reference model is settled to zero and the nose of the helicopter is shaken slightly with small variation, the position of actuators change in order to bring back the system into the still condition i.e. the MPC gives the action to the system.

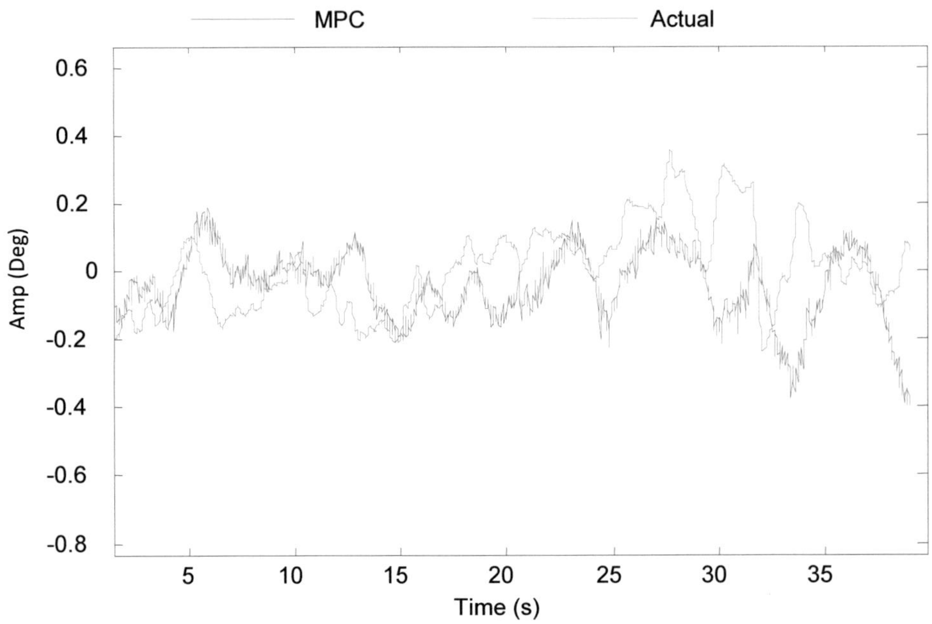

Fig. 7. generated lateral input (MPC) vs lateral command

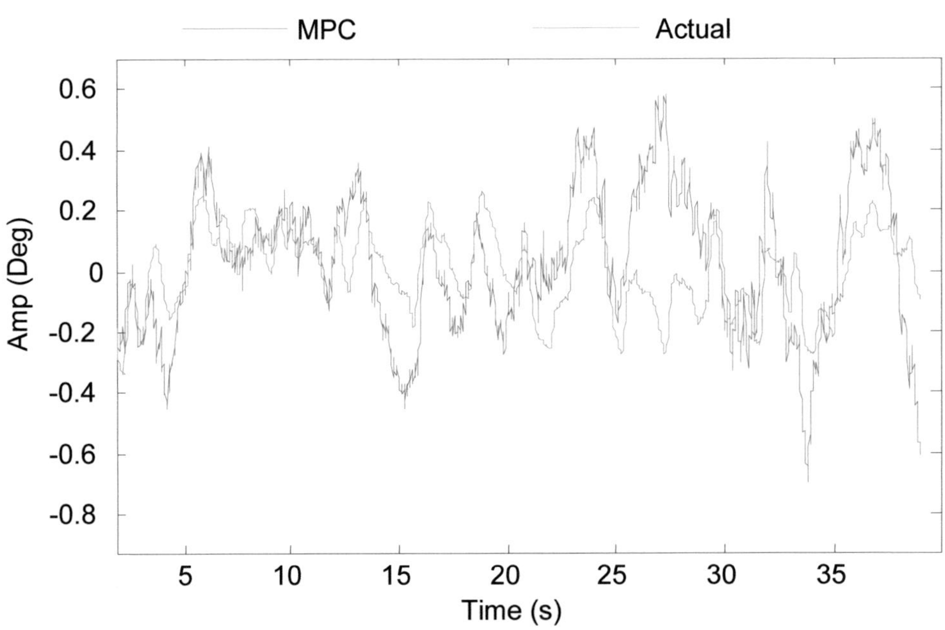

Fig. 8. generated longitudinal input (MPC) vs longitudinal command.

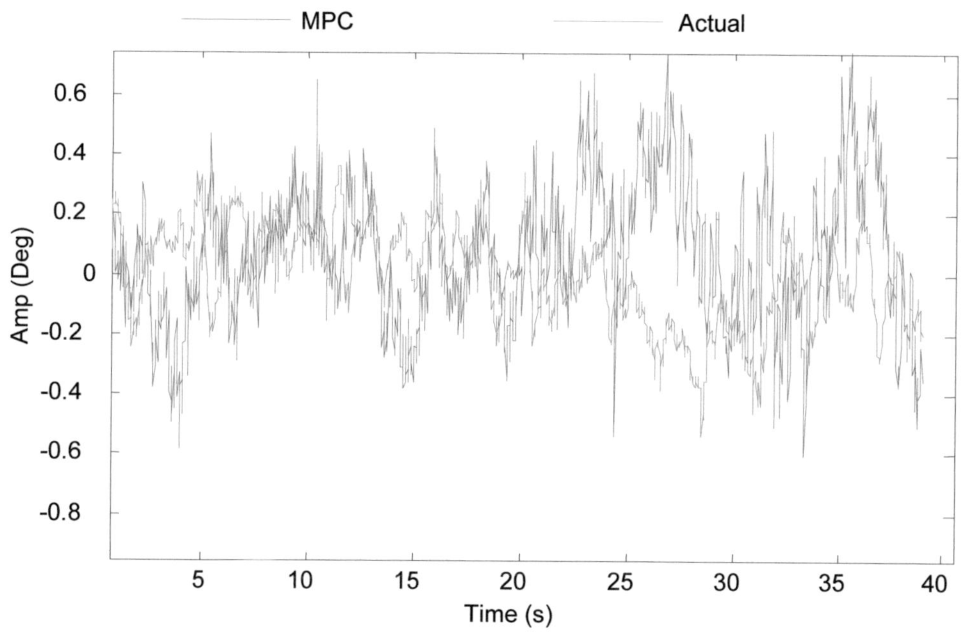

Fig. 9. generated pedal input by MPC vs pedal command.

6. Conclusion

In this paper, a MIMO model predictive control (MPC) system is implemented into hardware in the loop based xpc-target rapid-prototype system to guarantee the equilibrium of the helicopter platform. The MIMO MPC design was carried out using an experimentally estimated model of the Helicopter. The performance of the controller is tested in simulation and hardware in the loop using different set-point scenarios. Simulation results showed that the controller can efficiently stabilize the system under all the introduced disturbances. A real time controller based on xpc-target rapid prototype is developed to implement the proposed controller. The ground results proved that the proposed real time MPC can sufficiently stabilize the system in hovering conditions.

7. Acknowledgment

The authors gratefully acknowledge the support from MOSTI (Malaysia) Sciencefund: Hardware-in-the-Loop Simulation for Control System of Mini Scale Rotorcraft project No. 13-01-03-SF0024. The previous team researchers Mr Terran and KC Yap are gratefully acknowledged for their help.

8. References

Abdelhakim Deboucha, Zahari Taha, 2010. Identification and control of small-scale helicopter. *Journal of Zhejiang University Science A* (Applied Physics & Engineering). Vol. 11 (12), pp. 978-985

Abdelhakim Debooucha, 2011. Neural Network-based modelling and control design of small scale helicopter. *Master's Dissertation, University of Malaya*, Malaysia.

Cai, G., Peng, K., Chen, B. M., Lee T. H., (2005). Design and assembling of a UAV helicopter system. *IEEE Proceedings of the 5th International Conference on Control and Automation* (pp. 697–702), Hungary.

Del-Cerro, J., Barrientos. A., Campoy. P., & Garcia. P. (2002). An autonomous helicopter guided by computer vision for inspection of overhead power cable. *International Conference on Intelligent Robots and Systems* (pp. 69–78). Switzerland.

Doherty, P., Granlund, G., Kuchcinski, K., Sandewall, E., Nordberg, K., Skarman, E., Wiklund, J., (2000). The WITAS unmanned aerial vehicle project. Proceedings of the *14th European Conference on Artificial Intelligence* (ECAI) (pp. 747–755). Berlin, Germany.

Eck, C., Chapuis, J., & Geering, H. P. (2001). Inexpensive autopilots for small unmanned helicopters. *Proceedings of the Micro and Mini Aerial Vehicles Conference*. (pp. 1–8). Brussels, Belgium.

Edgar, N. S., Hector, M. B., Carlos, M. V. (2007). Combining fuzzy, PID and regulation control for an autonomous mini-helicopter. *Information Sciences*, 177(10), 1999–2022. doi:10.1016/j.ins.2006.10.001.

Garcia, R. D. & Valavanis, K. P., (2009). The implementation of an autonomous helicopter testbed. *Journal of Intelligent & Robotics Systems*, 54(3). doi: 10.1007/10846-008-9273

Jay, H. L., Manfred. M., Carlos. E. G. (1994). State space interpretation of model predictive control. *Automatica*, 30(4), 707–714. doi:10.1016/0005-1098(94)90159-7

Kadmiry, B., Driankov, D. (2004). A fuzzy flight controller combining linguistic and model-based fuzzy control. *Fuzzy Sets and Systems*,146, 313–347.

Kim, H. J., and Shim, D. H. (2003). A flight control system for aerial robots: Algorithms and experiments. *Control Engineering Practice*, 11, 1389– 1400

Liuping, W. (2008). Model predictive control system design and implementation using MATLAB. Melbourne, Australia: Springer.

Mario, G. P. (1997). A modified genetic algorithm for the design of autonomous helicopter control system. *American Institute of Aeronautics and Astronautics*, pp. 1–10.

Montgomery, J. F., & Bekey, G. A. (1998). Learning helicopter control through "teaching by showing". *In Proceedings of the 37th IEEE Conference on Decision and Control*. Tampa, Florida, USA.

Montgomery, J. F., Fagg, A. H., & Bekey, G. A. (1995). The USC AFV-I: A behaviour based entry. *1994 Aerial Robotics Competition IEEE Expert*, pp. 16–22.

Taha. Z., Tang. Y. R., Yap. K. C. (2010). Development of an onboard system for flight data collection of small- scale helicopter. *Mechatronics, Vol.No*, 1–13.

Wenbing, C., Xia, X., Ming, L., Yunjian, G., & Min, Z. (2007). Sensor based adaptive neural network control for small-scale unmanned helicopter. *International Conference on Information Acquisition* (pp. 187–191). Jeju City, Korea.

Srikanth, S., James, F. M., & Gaurav, S. S. (2003). Visually guided landing of an unmanned helicopter. *IEEE Transactions on Robotics and Automation*, 19(3), 371–381. doi: 10.1109/TRA.2003.810239.

Zahari, T., Yap. K. C., Tang, Y. R. (2008). Avionics box for small unmanned helicopter. *Proceedings of the 9th Asia Pacific Industrial Engineering & Management Systems Conference* (pp. 2841 - 2845). Bali, Indonesia.

Zhang, Q. & Ljung., (2004). Multiples steps prediction with nonlinear ARX models. Proceedings of the *6th international Federation of Automatic Control (IFAC). Nonlinear control systems* (pp. 309–314).

Adaptable PID Versus Smith Predictive Control Applied to an Electric Water Heater System

José António Barros Vieira[1] and Alexandre Manuel Mota[2]
[1]*Polytechnic Institute of Castelo Branco, School of Technology of Castelo Branco, Department of Electrical and Industrial Engineering,*
[2]*University of Aveiro, Department of Electronics Telecommunications and Informatics, Portugal*

1. Introduction

Industry control processes presents many challenging problems, including non-linear or variable linear dynamic behaviour, variable time delay that means time varying parameters. One of the alternatives to handle with time delay systems is to use prediction technique to compensate the negative influence of the time delay. Smith predictor control (SPC) is one of the simplest and most often used strategies to compensate time delay systems. In this algorithm it is important to choose the right model representation of the linear/non-linear system. The model should be accurate and robust for all working points, with a simple mathematical and transparent representation that makes it interpretable.

This work is based in a previews study made in modelling and controlling a gas water heater system. The problem was to control the output water temperature even with water flow, cold water temperature and desired hot water temperature changes. To succeed in this mission one non-linear model based Smith predictive controller was implemented. The main study was to identify the best and simple model of the gas water heater system.

It has been shown that many variable industry linear and non-linear processes are effectively modelled with neural and neuro-fuzzy models like the chemical processes (Tompson & Kramer, 1994). Hammerstein and Wiener models like pH-neutralization, heat exchangers and distillation columns (Pottman & Pearson, 1992), (Eskinat et al., 1991). And hybrid models like heating and cooling processes, fermentation (Psichogios & Ungar, 1992), solid drying processes (Cubillos et al., 1996) and continues stirred tank reactor (CSTR) (Abonyi et al., 2002).

In this previews work there were explored this three different modelling types: neuro-fuzzy (Vieira & Mota, 2003), Hammerstein (Vieira & Mota, 2004) and hybrid (Vieira & Mota, 2005) and (Vieira & Mota, 2004a) models that reflex the evolution of the knowledge about the first principles of the system. These kinds of models were used because the system had a non-linear actuator, time varying linear parameters and varying dead time systems. For dead time systems some other sophisticated solutions appear like in (Hao, Zouaoui, et al., 2011) that used a neuro-fuzzy compensator based in Smith predictive control to achieved better results. Or other solutions for unknown dead time delays like (Dong-Na, Guo, et al., 2008) that use gray predictive adaptive Smith-PID control because the dead time variation is unknown. There is an interesting solution to control processes with variable time delay using EPSAC (Extended Prediction Self-Adaptive Control) (Sbarciog, Keyser, et al., 2008) that could be used in this systems because the delay variations is caused by fluid transportation.

At the beginning there was no knowledge about the physical model and there were used black and grey box model approaches. Finally, the physical model was found and a much simple adaptive model was achieved (the physical model white box modelling).

This chapter presents two different control algorithms to control the output water temperature in an electric water heater system. The first approach is the adaptive proportional integral derivative controller and second is the Smith predictive controller based on the physical model of the system. From the previews work it is known that the first control approach is not the best algorithm to use in this system, it was used just because it has a simple mathematical structure and serves to compare results with the Smith predictive controller results. The Smith predictive controller has a much more complex mathematical structure because it uses three internal physical models (one inverse and two directs) and deals with the variable time delay of the system. The knowledge of the physical model permits varying the linear parameters correctly in time and gives an interpretable model that facilitate its integration on any control schemes.

This chapter starts, in section 2, with a full description of the implemented system to control the electric water heater, including a detailed description of the heater and its physical equations allowing the reader to have a comprehension of the control problems that will be explained in later sections.

Section 3 and 4, describes the two control algorithms presented: the adaptive proportional integral derivative control structure and the Smith predictive control based in the physical models of the heater. These sections show the control results using the two approaches applied in to a domestic electric water heater system. Finally, in section 5, the conclusions are presented.

2. The electric water heater

The overall system has three main blocks: the electric water heater, a micro-controller board and a personal computer (see figure 1).

The micro-controller board has two modules controlled by a flash-type micro-controller from the ATMEL, ATMEGA168 with 8Kbytes on FLASH. The interface module has the necessary electronics to connect the sensors and control the actuator. The communication module has the RS232 interface used for monitoring and acquisition of all system variables in to a personal computer.

After this small description of the prototype system, the electric water heater characteristics are presented and its first principles equations are presented.

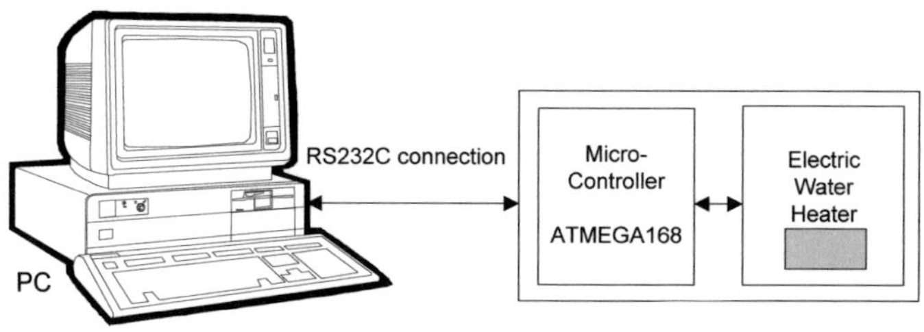

Fig. 1. System main blocks.

2.1 Electric water heater description

The electric water heater is a multiple input single output (MISO) system. The controlled output water temperature will be called hot water temperature (*hwt(t)*). This variable depends of the cold water temperature (*cwt(t)*), water flow (*wf(t)*), power (*p(t)*) and of the electric water heater dynamics. The hot and cold water temperature difference is called delta water temperature ($\Delta t(t)$).

The electric water heater is physically composed by an electric resistance, a permutation chamber and several sensors used for control and security of the system as shown on figure 2.

Operating range of the *hwt(t)* is from 20 to 50ºC. Operating range of the *cwt(t)* is from 5 to 25ºC. Operating range of the *wf(t)* is from 0,5 to 2,5 litters / minute. Operating range of the *p(t)* is from 0 to 100% of the available power.

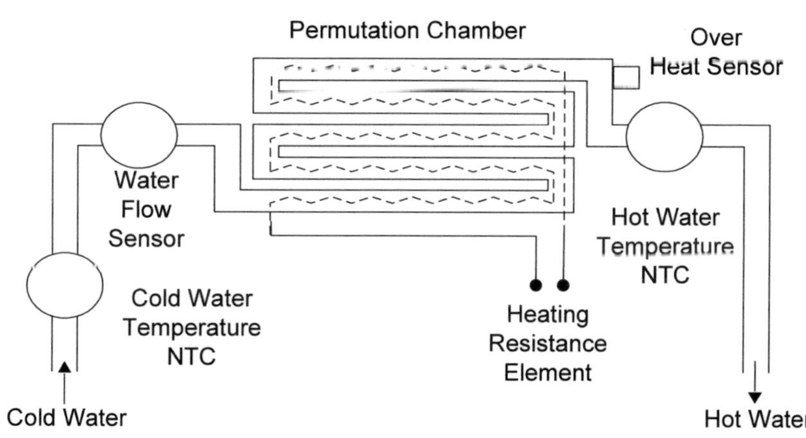

Fig. 2. Schematic of the electric water heater: sensors and actuator.

The applied energy in to the heating resistance is controlled using 100 alternated voltage cycles (one second). In each sample, the applied number of cycles is proportional to the delivery energy to the heating element.

Figure 3 shows one photo of the electric water heater and the micro-controller board.

Fig. 3. Photo of the electric water heater and the micro-controller board.

2.2 Electric water heater first principles equations

Applying the principle of energy conservation in the electric water heater system, equation 1 could be written. This equation was based in a previews work made in modelling a gas water heater system, first time presented in [11].

$$\frac{dEs(t)}{dt} = Qe(t - td) - wf(t)hwt(t)Ce - wf(t)cwt(t)Ce \tag{1}$$

Where $dEs(t)/dt = MCe(d\Delta t(t)/dt)$ is the energy variation of the system in the instant t, $Qe(t)$ is the calorific absorbed energy, $wf(t)cwt(t)Ce$ is the input water energy that enters in the system, $wf(t)hwt(t)Ce$ is the output water energy that leaves the system, and Ce is the specific heat of the water, M is the water mass inside of the permutation chamber and td is the variable system time delay.

The time delay of the system has two parts: a fixed one that became from the transformation of energy and a variable part that became from the water flow that circulates in the permutation chamber.

M is the mass of water inside of the permutation chamber (measured value of 0,09Kg) and *Ce* is the specific heat of the water (tabled value of 4186 J/(KgK)). The maximum calorific absorbed energy *Qe(t)* is proportional to the maximum electric applied power of 5,0 KW.

The absorbed energy *Qe(t)* is proportional to the applied electric power *p(t)*. On each utilization of the water heater it was considered that *cwt(t)* is constant, it could change from utilization to utilization, but in each utilization it remains approximately constant. Its dynamics does not affect the dynamics of the output energy variation because its variation is too slow.

Writing equation 1 in to the Laplace domain and considering a fixed water flow $wf(t)=Wf$ and fixed time delay td, it gives equation 2.

$$\frac{\Delta t(s)}{Qe(s)} = \frac{\frac{1}{WfCe}}{\frac{M}{Wf}s+1}e^{-s\,td} = \frac{\frac{1}{WfCe}\frac{Wf}{M}}{s+\frac{Wf}{M}}e^{-s\,td} \tag{2}$$

Passing to the discrete domain, with a sampling period of h=1 second and with discrete time delay $\tau d(k) = \mathrm{int}(\frac{td(t)}{h})+1$, the final discrete transfer function is illustrated in equation 3.

$$\Delta t(k+1) = \left(e^{-\frac{Wf}{M}}\right)\Delta t(k) + \left(\frac{1}{WfCe}\left(1-e^{-\frac{Wf}{M}}\right)\right)\left(Qe(k-\tau d(k))\right) \tag{3}$$

The real discrete time delay $\tau d(k) = \tau d_1(k) + \tau d_2(k)$ is given in equation 4, where $\tau d_1(k) = 3s$ is the fixed part of $\tau d(k)$ that became from the transformation of energy $\tau d_2(k)$ and is the variable part of $\tau d(k)$ that became from the water flow *wf(k)* that circulates in the permutation chamber.

$$\tau d(k) = \begin{cases} 4 & to \quad wf(k) >= 1{,}75l/\min \\ 5 & to \quad 1{,}00l/\min < wf(k) < 1{,}75l/\min \\ 6 & to \quad wf(k) <= 1{,}00l/\min \end{cases} \tag{4}$$

Considering now the possibility of changes in the water flow, in the discrete domain *Wf=wf(k)* and $\tau d_2(k)$, the final transfer function is given in equation 5.

$$\begin{aligned}\Delta t(k+1) = {} & \left(e^{-\frac{wf(k-\tau d_2(k))}{M}}\right)\Delta t(k) + \\ & \left(\frac{1}{wf(k-\tau d_2(k))Ce}\left(1-e^{-\frac{wf(k-\tau d_2(k))}{M}}\right)\right)\left(Qe(k-\tau d(k))\right)\end{aligned} \tag{5}$$

Observing the real data of the system, the absorbed energy *Qe(t)* is a linear static function *f(.)* proportional to the applied electric power *p(t)* as expressed in equation 6.

$$Qe(k-\tau d(k)) = f\left(p(k-\tau d(k))\right) \tag{6}$$

Finally, the discrete global transfer function is given by equation 7.

$$\begin{aligned}\Delta t(k+1) = {} & \left(e^{-\frac{wf(k-\tau d_2(k))}{M}}\right)\Delta t(k) + \\ & \left(\frac{1}{wf(k-\tau d_2(k))Ce}\left(1-e^{-\frac{wf(k-\tau d_2(k))}{M}}\right)\right)\left(f\left(p(k-\tau d(k))\right)\right)\end{aligned} \tag{7}$$

If $A(k)$ and $B(k)$ are defined as expressed in equation 8, the final discrete transfer function is given as defined in equation 9.

$$A(k) = e^{-\frac{wf(k-\tau d_2(k))}{M}}$$
$$B(k) = \frac{1}{wf(k-\tau d_2(k))Ce}\left(1 - e^{-\frac{wf(k-\tau d_2(k))}{M}}\right) \tag{8}$$

$$\Delta t(k+1) = A(k)\Delta t(k) + B(k)\left(f\left(p(k-\tau d(k))\right)\right) \tag{9}$$

2.3 Physical model validation

For validation of the presented discrete physical model, it is necessary to have open loop data of the real system. This data has been chosen to respect two important requirements: frequency and amplitude spectrum wide enough (Psichogios & Ungar, 1992). Respecting the necessary presupposes, the collect data is made via RS232 connection to the PC. The validation data and the physical model error are illustrated in figure 4.

Figure 4 shows the physical model error signal *e(k)*, which is equal to the difference between delta and estimated delta water temperature *e(k)= Δt(k)- Δtestimated(k)*. It can be seen from this signal, that the proposed model achieved very good results with a mean square error (MSE) of 1,32°C^2 for the all test set (1 to 1600).

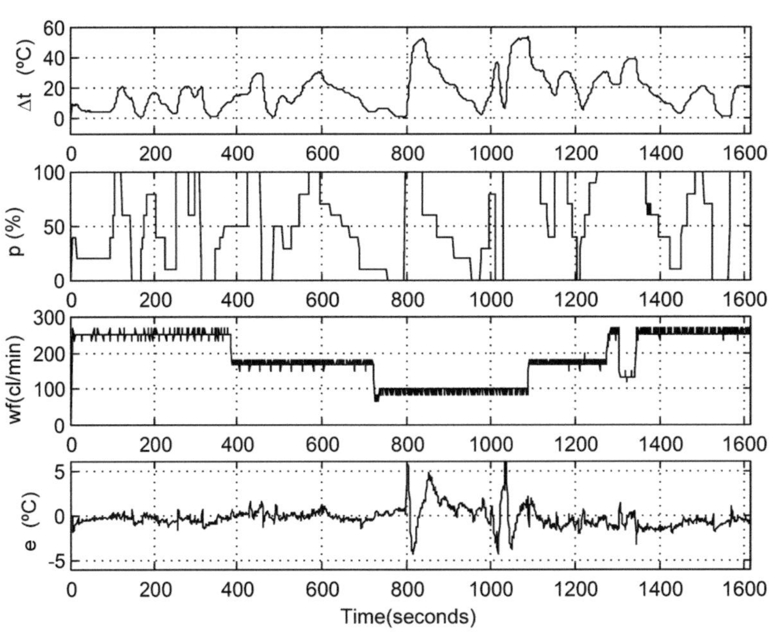

Fig. 4. Open loop data used to validate the model.

From the validation test, figure 5 shows the two linear variable parameters expressed in equation 8 of the physical model used.

As can be seen the *A(k)* parameter that multiply with the regressor delta water temperature changes significantly with water flow *wf(k)* and the B(k) parameter that multiply with the regressor applied power $f\left(p(k-\tau d(k))\right)$ presents very small changes with the water flow *wf(k)*.

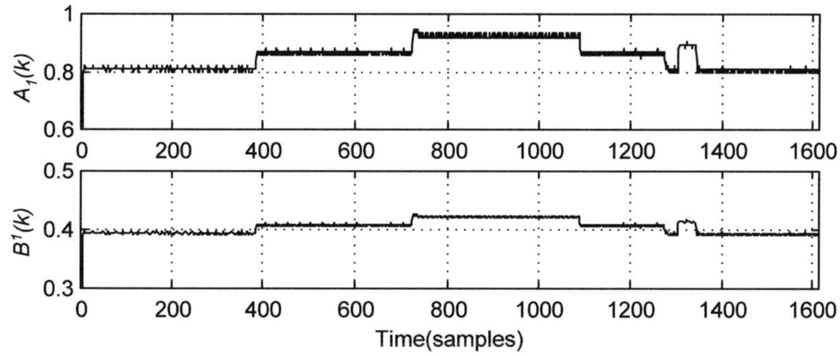

Fig. 5. The two linear variable parameters *A(k)* and *B(k)*.

From the results it can be seen that for the small water flows the model presents a bigger error signal. This happens because of the small resolution of the water flow measurements and of the estimated integer time delays forced (a multiple of the sampling time h it is not possible fractional time delays).

3. Adaptive PID controller

The first control loop tested is the adaptive proportional integral derivative control algorithm. Adaptive because we know that gain and time constant of the system changes with the input water flow. First it is described the control structure and its parameters and second the real control results are showed.

3.1 Adaptive PID control structure

This is a very simple and well known control strategy that has two control parameters *Kp* and *Kd* that are multiplied by the water flow, as illustrated in figure 6. The applied control signal is expressed in equation 10:

$$\begin{aligned} f\left(p(k)\right) &= f\left(p(k-1)\right) + wf(k)K_p e(k) \\ &+ wf(k)K_d(e(k)-e(k-1)) \end{aligned} \tag{10}$$

The P block gives the error proportional contribution, the *D* block gives the error derivative contribution and the *I* block gives the control signal integral contribution.

The three control parameters were adjusted after several experimental tests in controlling the real system. This algorithm has some problems dealing with time constant and time delay variations of the system. With this control loop it is not possible to define a close loop

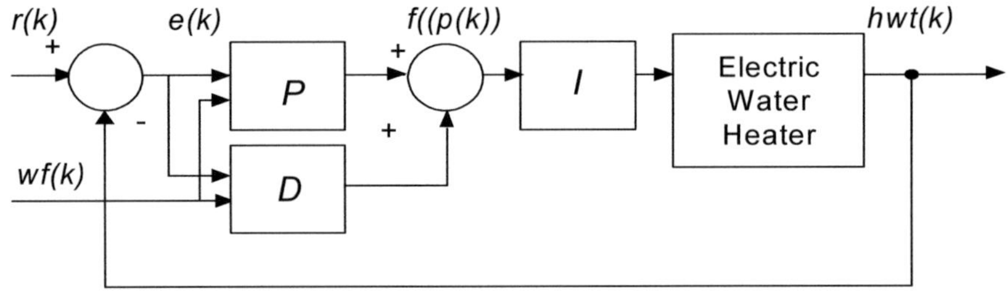

Fig. 6. APID controller constituent blocks.

system with a fixed time constant. The time delay is also a problem that is not solved with this control algorithm.

It was define a reference signal *r(t)* that is the desired hot water temperature and a water flow *wf(t)* with several step variations similar to the ones used in real applications. The cold water temperature was almost constant around 13,0 ºC.

For testing the controllers it can be seen that error signal *e(t)=r(t)-hwt(t)* is around zero excepted in the input transitions. In reference step variations it can be seen that the overshoots for the different water flows are similar but the rise times are clearly different, for small water flows the controller presets bigger rise times. In water flow variations the control loop have some problems because of the variable time delay. This control loop only reacted when error appears.

3.2 Adaptive PID control results

With the proposed tests signals, the tuned adaptive PID control structure was tested in controlling the electric water heater. The APID control results are shown in figure 7.

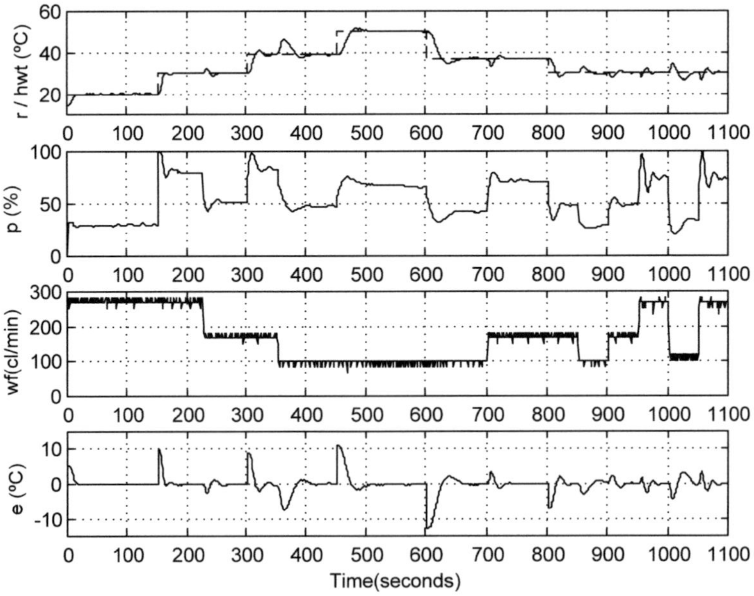

Fig. 7. Adaptive PID control results.

As it was predicted the results have shown some problems in water flow variations because the controller just reacts when it feels an error signal different from zero.

The evaluation control criterion used is the mean square error (MSE). The MSE in the all test is presented in table 1.

Algorithm	MSE Test Set
APID	5,97

Table 1. Mean square errors of the control results.

4. Smith predictive controller

The second control loop tested is the Smith predictive control algorithm. This control strategy is particularly used to control systems with time delay. First it is described the control structure and its parameters and second the control results are showed.

4.1 Smith predictive control structure

The Smith predictive controller is based in the internal model controller architecture that uses the physical model presented in section II, as illustrated in figure 8. It uses two physical direct models one with time delay for the prediction loop and another with out the time delay for the internal model control structure.

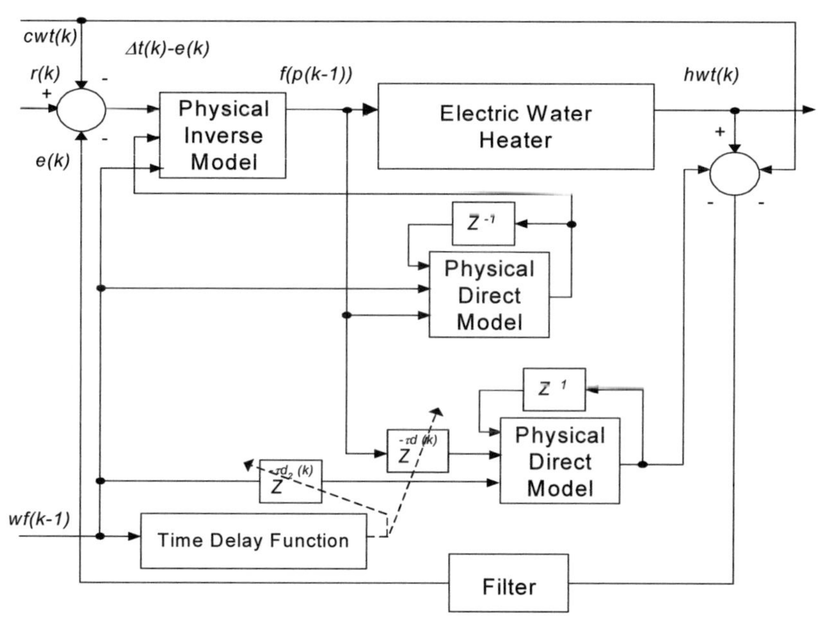

Fig. 8. SPC constituent blocks.

The Smith predictive control structure has a special configuration, because the systems has two inputs with two deferent time delays so it uses two direct models, one model with time delay for compensate its negative effect and another with out time delay needed for the internal model control structure.

The SPC separates the time delay of the plant from time delay of the model, so it is possible to predict the *Δt(k), τd(k)) steps* earlier, avoiding the negative effect of the time-delay in the control results. The time delay is a known function that depends of the water flow *wf(k)*. The incorrect prediction of the time delay may lead to aggressive control if the time delay is under estimated or conservative control if the time delay is over estimated (Tan & Nazmul Karim, 2002), (Tan & Cauwenberghe, 1999).

The physical inverse model is mathematically calculated based in the physical direct model presented in section 2 used with out time delay.

The low pass filter used in the error feedback loop is a digital first order filter used to filter the feedback error and indirectly to filter the control signal *f(p(k))*. The time delay function is a function of the water flow, which is explained in section 2 and expressed in equation 4.

To test the SPC based in the physical model it was used the same reference signals *r(t)* and water flow *wf(t)* used to test the adaptive PID controller.

4.2 Smith predictive control results

The SPC results are shown in figure 9. As it was predicted from previews work the results are very good in reference and in water flow changes. The behaviour of the closed loop system is very similar in every working point.

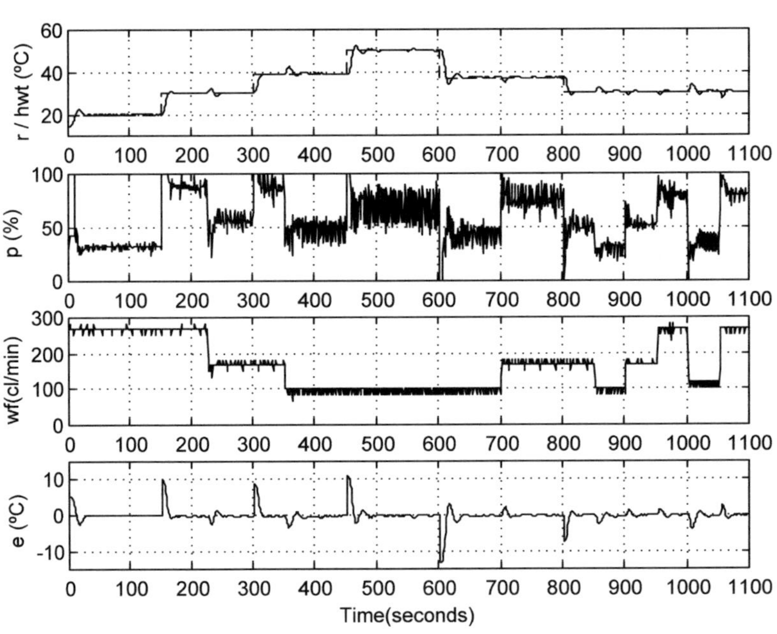

Fig. 9. SPC control results.

It can be seen that for small water flows the resolution of the measure is small that makes the control signal a bit aggressive but it does not affect the output hot water temperature.

For small water flows there is another problem with the multiplicity of the time delay and its resolution. With a sampling period of 1 second it is more difficult to use factional time delays that happen in reality. This makes the control results a bit aggressive.

The final MSE evaluation control criterion achieved with the SPC is presented in table 2.

Algorithm	MSE Test Set
SPC	3,56

Table 2. Mean square errors of the control results.

The physical model includes à priori knowledge of the real system and has the advantage of been interpretable. This characteristic facilitates the implementation and simplicity the Smith predictive control algorithm.

5. Conclusions

For comparing the two control algorithms, APID and SPC, the reference signals were applied in controlling the system and the respective mean square errors were calculated as showed in table 1 and 2.

This work present and validate the physical model of the electric water heater. This model was based in the model of a gas water heater because of the similarities of both processes.

The MSE of the validation test is very small which validate the physical electric water heater model accuracy.

Finally, the proposed APID and SPC controllers were successful applied in the electric water heater system. It is verify that the SPC achieved much better results than the adaptive proportional integral derivative controller did as it was expected because of the system characteristics.

The best control structure for varying first order systems with varying large time delay is the Smith predictive controller based in physical model of the system as presented in this work. The SPC controller proposed in opposition to the APID controller reacts also very well in cold water temperature variations.

This controller is mathematically simple and easily implemented in a microcontroller with reduce resources.

For future work some improvements should be made as the enlargement of the resolution of the used water flow and the redefinition of the time delay function.

6. References

Tompson M. L. and Kramer M. A., (1994). Modelling chemical processes using prior knowledge and neural networks, *A. I. Ch. E. Journal*, 1994, vol. 40(8), pp. 1328-1340.

Pottman M., Pearson R. K., (1998). Block-Oriented NARMAX Models with Output Multiplicities, *AIChE Journal*, 1998, vol. 44(1), pp. 131-140.

Eskinat E., Johnson S. H. and Luyben W., (1991). Use of Hammerstein Models in Identification of Non-Linear Systems, *AIChE Journal*, 1991, vol. 37(2), pp. 255-268.

Psichogios D. C. and Ungar L. H., (1992). A hybrid neural network-first principles approach to process modelling, AIChE Journal, 1992, vol. 38(10), pp. 1499-1511.

Cubillos F. A., Alvarez P. I., Pinto J. C., Lima E. L., (1996). Hybrid-neural modelling for particulate solid drying processes. *Power Thecnology*, 1996, vol. 87, pp. 153-160.

Abonyi J., Madar J. and Szeifert F., (2002). Combining First Principles Models and Neural Networks for Generic Model Control, *Soft Computing in Industrial Applications - Recent Advances*, Eds. R. Roy, M. Koppen, S. O., T. F., F. Homann Springer Engineering Series, 2002, pp.111-122.

Vieira J., Mota A. (2003). Smith Predictor Based Neural-Fuzzy Controller Applied in a Water Gas Heater that Presents a Large Time-Delay and Load Disturbances, *Proceedings IEEE International Conference on Control Applications*, Istanbul, Turkey, 23 a 25 June 2003, vol. 1, pp. 362-367.

Vieira J., Mota A. (2004). Parameter Estimation of Non-Linear Systems With Hammerstein Models Using Neuro-Fuzzy and Polynomial Approximation Approaches, *Proceedings of IEEE-FUZZ International Conference on Fuzzy Systems*, Budapest, Hungary, 25 a 29 July 2004, vol. 2, pp. 849-854.

Vieira J., Dias F. and Mota A. (2005). Hybrid Neuro-Fuzzy Network-Priori Knowledge Model in Temperature Control of a Gas Water Heater System, *Proceedings of 5th International Conference on Hybrid, Intelligent Systems*, Rio de Janeiro, 2005.

Vieira J. and Mota A. (2004a). Water Gas Heater Non-Linear Physical Model: Optimization with Genetic Algorithms. *Proceedings of IASTED 23rd International Conference on Modelling, Identification and Control*, February 23-25, vol. 1, pp. 122-127.

Psichogios D. C. and Ungar L. H. (1992) A hybrid neural network-first principles approach to process modelling, *Journal AIChE*, vol. 38(10), pp. 1499-1511.

Tan Y. and Nazmul Karim M. (2002) Smith Predictor Based Neural Controller with Time Delay Estimation. *Proceedings of 15th Triennial World Congress*, IFAC.

Tan Y. and Van Cauwenberghe A. R. (1999) Neural-network-Based d-step-ahead Predictors for Nonlinear Systems with Time Delay. Engineering Applications of Artificial Intelligence. Vol. 12(1), pp. 21-35.

Hao Chen, Zouaoui, Z. and Zheng Chen (2011) A neuro-fuzzy compensator based Smith predictive control for FOPLDT process. *Proceedings of International Conference on Mechatronics and Automation (ICMA)*, 2011, pp. 1833 – 1838.

Dong-Na Shi, Guo Peng and Teng-Fei Li (2008) Gray predictive adaptive Smith-PID control and its application. *Proceedings of International Conference on Machine Learning and Cybernetics*, 2008, vol. 4, pp. 1980 – 1984.

M. Sbarciog, R. De Keyser, S. Cristea and C. De Prada (2008) Nonlinear predictive control of processes with variable time delay, A temperature control case study. *Proceedings of IEEE Multi-conference on Systems and Control Applications*, San Antonio, Texas, USA, September, 2008, pp. 3-5.